ENVIRONMENTAL QUALITY IN OFFICES

ENVIRONMENTAL QUALITY IN OFFICES

Jacqueline C. Vischer

With a foreword by the president of NAHB National Research Center

Library of Congress Catalog Card Number 88-17116

ISBN 0-442-20498-1

Printed in the United States of America

Text Design by Monika Grejniec

Van Nostrand Reinhold
115 Fifth Avenue
New York, New York 10003

Van Nostrand Reinhold (International) Limited
11 New Fetter Lane
London EC4P 4EE, England

Van Nostrand Reinhold
480 La Trobe Street
Melbourne, Victoria 3000, Australia

Macmillan of Canada
Division of Canada Publishing Corporation
164 Commander Boulevard
Agincourt, Ontario M1S 3C7, Canada

16 15 14 13 12 11 10 9 8 7 6 5 4 3 2 1

Library of Congress Cataloging in Publication Data

Vischer, Jacqueline.
Environmental quality in offices.

Bibliography: p.
Includes index.
1. Work environment. 2. Offices. 3. Buildings—Environmental engineering. I. Title.
T59.77.V57 1989 725′.23 88-17116
ISBN 0-442-20498-1

Dedicated to the memory of
Professor Roslyn Lindheim
1922–1987

CONTENTS

FOREWORD

It is truly a joy to encounter a book that comprehensively covers and clearly explains the complex subject of the influence the physical environment has on people in the work place. After discussing the general topic of environmental quality in offices, Dr. Vischer delineates the building-in-use assessment system, an extremely useful and unobtrusive way to help manage people in their work environment and ensure that the environment is productive and comfortable.

This book will appeal to managers of office buildings, students of architectural psychology, sociologists interested in methodological issues, building products designers, and participants in standards-setting activities. Each person who reads this book will do so with a unique set of problems and concerns as well as knowledge; thus the information in *Environmental Quality in Offices* also will be used very differently by each of us.

The discussions both of techniques for building performance measurement and of environmental psychology are useful for those involved in development, design, and construction, or in managing and operating large complex buildings. Discussions of modern office building technology and office design philosophies will assist the reader in understanding appropriate courses of action to remedy assessed weaknesses. The strong theoretical underpinnings of the book, which provide a basis for the assessment system in the empirical research tradition, provide informative and provocative reading to anyone interested in building science, design research, and the future of the building industry. Dr. Vischer's book will be of interest to all serious students of the human and organizational impact of our buildings.

My personal interest in this book is threefold. First, as a manager responsible for the productivity of a staff of highly trained senior professionals, I have been helped by *Environmental Quality in Offices* to cope with the many daily decisions about space and location, as well as to respond to complaints about environmental quality. I can see applying the building-in-use assessment system to our new Research and Development Center to objectively monitor the environment, assess problems, and improve the building. Second, as a student of the influence of the built environment on individuals and organizations for the past twenty-five years, I have often had to make sense of the frequently disparate literature and research in this field. This clear, concise book has proved to be a help in this task. Third, as a professional committed to the improvement of the building industry, I see this book making the critical contribution of presenting a long-needed tool—the building-in-use assessment system—for assessing and improving our product's performance.

Feedback to building managers and owners presently consists of workers' complaints and the reactions of individual supervisors. Personal complaints may simply be dismissed or be responded to because the person complaining is a prominent individual. The building-in-use assessment system, described in detail in this book, is structured to collect data in ways that prevent workers' anxieties and managers' preconceptions from tainting the process of building improvement. The results of such assessments will be particularly useful when employed together with the information in Chapter 8 describing how the physical environment may be modified to address and respond to problems identified by the assessment.

Environmental Quality in Offices also sends an important message to the administrators of codes and standards who deal with the quality of indoor environments: mere compliance with standards does not always get builders the environmental quality they want. Our present codes and standards rely heavily on what we believe to be our ability to measure and standardize physical attributes. This book, especially Chapter 6, "The Building-in-Use Dimensions of Environmental Quality," provides a critical missing ingredient: information on how environments are experienced by those who work in them. Such feedback can dramatically improve the linked processes and diversity of interests represented in applying codes and standards to the design, construction, and operation of our buildings.

Much of previous building research has assessed office environments in order to prescribe better design or design approaches. In the approach presented here, the users of the data decide how they want to respond to their findings in the context of their particular resources and unique goals. It is refreshing to find not prescriptions, but rather a way to make wiser decisions.

Practical application of the building-in-use assessment system will enable building managers and owners to improve existing buildings, as well as encourage leading manufacturers and builders to invent new, more effective products. Application might involve the following steps:

1. Listen for complaints;
2. Respond to complaints with the building-in-use questionnaire;
3. Use the data to correct the problem in the building as efficiently as possible;
4. Modify the products on the basis of the findings to avoid problems in future buildings.

Following these steps will also lead to other intriguing areas of investigation, for example, the comparative results of applying the building-in-use assessment system to a wide range of work spaces, such as government office buildings, industrial research laboratories, private corporations, and the home office, to name just a few. Will we find consistency or diversity? There are many new things can we learn about the different locations in which we choose to work?

David J. MacFadyen, President
National Association of Home Builders (NAHB)
National Research Center
Upper Marlboro, Maryland

PREFACE

A few miles north of Boston, in the suburbs, where new modern office buildings are bursting like locusts into the rolling verdant landscape, a handsome new red-brick complex hovering over the freeway advertises new space for lease and proclaims to passing drivers on a huge black and white banner across its upper story, "Because Doing Business Is Enough of a Struggle."

This is an extraordinary proclamation. This building is saying that in and of itself, it has something to offer. It is not selling a nice setting, a view, or modern electronic installations; it is selling the idea of office space as a treat and a relief, as an active participant in daily life at work. This is a new idea.

Most office buildings have been built and occupied as static, incidental, and neutral shelters for people's tasks. Some office spaces are better than others, but for many employers and employees, the office is an inert, dull background to the dynamic ups and downs of their working life.

This book maintains that all office space could improve if it were viewed the same way that the Boston building proclaims itself. To view the office building as dull, static, and a necessary evil—even if it is a well-designed building—is the wrong approach. On top of this, offices are often badly designed and built. I have heard people say they left their jobs because they could not stand working in the building; or that they would never work downtown because they hate fluorescent lights and open-plan desk arrangements. Working at home is an integral part of the era of the electronic office. Many office workers take terminals home, and companies contract out to home-based workers. In some quarters there have been mutterings about the obsolescence of modern office buildings, and predictions that they will soon become mere shells void of all life except for janitors sweeping empty hallways, and elevators standing with their doors permanently open.

If there is any validity to this scenario, now is the time to change the image of the modern office. Most North American cities have experienced an office-building boom in recent years. Cities and their suburbs are full of large and small, high and low, electronically cabled, mechanically ventilated, brick, steel, glass, and concrete office buildings that together constitute an investment of capital and of our society's resources that our economy cannot afford to waste. The building that shouts to freeway drivers that it is a member of the team, an asset to the corporation, a reward, and an active contributor to working life is proclaiming an idea whose time has come. Its sign welcomes people and companies to the reality of the beautiful building, the office that makes life easier instead of more difficult, the office that actually helps people to do their work.

The time has come to think of the office building as an asset and an aid, and as a contributor to the effectiveness of employees. For consumers of office buildings, the time has come to be selective, to demand a better quality product, to ensure that the space *you* choose works for and contributes to the quality of *your* working life.

We realize that few companies are in a position to abandon their unsatisfactory offices immediately and look around for new space that meets new criteria of what an office ought to be doing for them. In this book we propose a way of making the office you have work properly for you. This is not just a question of implementing small and simple—or even large and complicated—environmental changes. It is a question of understanding which environmental improvements to implement and when to implement them, regardless of the scale of improvement. Some offices may be reasonably satisfactory; others may be awful: all could be improved.

In chapters 1 and 2 we review some of the reasons why office buildings do not work well for people. Some histories of modern office buildings tell sad and revealing stories about what is wrong with modern offices. It is important to understand that organizations have traditionally been reluctant to make major alterations to their office space, unless workers' health has been at risk or operating the building is consuming excessive amounts of money. In these chapters, environmental quality is discussed in terms of what it means and when and how it is recognized by building managers.

Scientists have studied office buildings, their technical performance, and the ways in which people use them, and chapters 3 and 4 are concerned with explaining what they have found and discussing the state of our knowledge. They explore the limitations of the psychological approach and of the technical or "building performance" approach to solving problems of users. It is clear that this knowledge, although not exhaustive, is not widely applied to the improvement of offices, either by the industry or by consumers (the users of office buildings). This book explores why the gains are so modest and outlines in some detail another way of looking not just at the study of buildings but at the way buildings and organizations work together and operate together as a system. The key to the rest of the book is in understanding that there is a new way of looking at the relationship between buildings and consumers.

Chapters 5, 6, and 7 carry this new way of looking at the problem through to details of the building-in-use assessment system that can be implemented by any office building owner or occupant with ease. The assessment system is a practical and systematic way of determining which environmental improvements to effect, and when to effect them, for *your* organization. Though not complicated to execute, the building-in-use assessment system is a way of turning an ordinary, "necessary evil," fair-to-middling office into a member of the work team, an active contributor, an asset, and a treat.

The final chapter of the book lists ways of solving environmental problems in offices, from improving the lighting to responding to workers' complaints about thermal comfort, to relieving stress from poor acoustics, to reducing indoor air pollution. Once the assessment system has told you what to improve, and how soon to do it, these are some of the actions that you can take to improve office quality.

The consumers of office buildings are not just the nine-to-five workers who put time in for a paycheck; they include corporate managers and senior executives, developers and realtors, highly paid technical and professional experts, janitors, maintenance staff, and facilities managers. They even include members of the public who do not work in the building but sit for lunch in its atrium, or visit its street-level shops and stroll across its plazas. Given the huge numbers of office buildings in our cities, and all the others that are still under construction, consumers have a right to a higher level of environmental quality than they currently receive. A lot of other things would have to happen before office environmental quality is assured; this book is about some of the first steps that consumers can take to make their office work better for them. Remember, doing business is enough of a struggle!

ACKNOWLEDGMENTS

The work on which this book is based was carried out by the team at Architectural and Building Sciences, Public Works Canada, Ottawa, directed by Jim Davison, and later, as the Building Performance Division, by William Cunningham. I would like to acknowledge their skill, energy, and team spirit and to thank them for making me part of their projects during those years.

There are many other individuals at PWC I would like to thank individually, but I will content myself with thanking them as a group, especially those who, through their own integrity, worked to help me clear my name and prove the value of my work against malicious accusations, and without whom this book could not have been written.

I owe special thanks to Richard Dillon, coauthor of the technical reports on which this book is partly based. His statistical expertise was invaluable, and he provided a constant stream of good, practical advice on the data analysis.

I would like to acknowledge the help of Harvey Bryan, Peter Ellis, Carl Rosenberg, and Ewart (Red) Wetherill—all busy experts in their own fields who took time to review parts of my manuscript and give me good advice. The illustrations were drawn by Boyd Rourke.

The ground was laid and the context established for this work and development of the building-in-use assessment system by John Zeisel, through his own consulting work for the government of Canada, and the generous sharing of his own ideas. I am most grateful for his many patient editings of the manuscript.

The opinions and ideas expressed in this book are my own and are in no way attributable to Public Works Canada.

Chapter 1

WHAT IS AN OFFICE BUILDING?

Some thirty million adults go to work every day in office buildings in North America. For white-collar workers who work an average productive life span, this means that about one third of their lives between the ages of twenty and sixty is spent in these environments. For people living in cities, regardless of where they work, the changing skyline and burgeoning takeover of downtown neighborhoods all over the world by towering high-rise office buildings is more and more a part of everyday life.

Tall and thin, or low and wide; white and shimmering, or black and forbidding; mirrored in silver or bronze or glass and steel; squared off at the top, or pointed to the sky like a finger of truth, or angled, as many are now, to collect energy from the sun: every one of us sees them, walks among them, lives with them, and often works in them (fig. 1-1).

1-1. A typical city skyline.

WHAT IS AN OFFICE BUILDING?

But what kind of places are these that we of the late twentieth century are accepting so readily into our lives? From the outside we know they change the skyline, create windy streets, block the sun, and occasionally provide passers by with open plazas or enclosed atria where they may sit and stroll. But what do we know of the insides, especially those of us who go to work in them every day of our lives?

This is a book about the insides of large office buildings and the experience of working in them. The buildings we see, enter, live and work in, look at, react to, love, and hate are an important part of our total environment. We tend to take them for granted. We tend to believe that building quality is the domain of the architect. Although everyone lives, works, and recreates in buildings, we, the people, the consumers of buildings, tend to take whatever we get. We may not like the look of a new office tower, or the design of the apartment we live in, or the acoustics of the theater we frequent, but we do not consider these feelings profound, and we rarely if ever act on them. Buildings are just there. We are not nearly as complacent about the quality of our *natural* environment, however. Consumer groups form and people get involved in protecting the quality of water, in retaining trees on building sites, in protecting fish and wildlife, in fighting for environmental quality.

The buildings we spend most of our time in are an equally demanding candidate for our attention. The quality of the buildings we occupy is just as important as that of the natural environment. As consumers of buildings, we have as much of a responsibility as the architect or any other building professional to ensure that standards of quality are maintained in the buildings we use and inhabit. Just as a building's form, colors, massing, and siting make up part of our experience of a city, so being inside a building fills out and forms our impressions of the company (in an office building), of the family (in a house), of the criminal justice system (in a prison), or of patient care (in a hospital). Increasing our awareness of the quality of the buildings we occupy increases our awareness that the environmental standards of these interiors often leave much to be desired. This is especially true of modern office buildings: the multistory, deep-interior-floor, mechanically ventilated, large-windowed variety in which most of the labor force currently spends most of its waking time. What is an office building? It is the modern equivalent of the ill-lit, unventilated, unsanitary factory settings in which most of the labor force worked in the early part of this century.

PROBLEMS IN OFFICE BUILDINGS

In most major cities architects and designers have designed office buildings whose shapes, forms, and colors look glamorous among the towers of midtown Manhattan, Chicago's loop, or downtown Houston. But what of the office spaces themselves? How many office building interiors are glamorous, colorful, or innovative? Relative to the number of new office buildings in North American cities, the amount of design innovation has been small.

From the inside, most of the glamorous postmodern office buildings of recent architectural vintage look the same as the ordinary cheap towers that are built by developers and are plentiful in the towns and suburbs of North America.

Inside the office buildings there are often large open-plan spaces and some enclosed offices for managers (fig. 1-2). On the ceiling, fluorescent lighting is lined up in rows. Sometimes there is task lighting, uplighting, or some other lighting innovation, but this is not routine. Desks and chairs are surrounded by acoustic panels or screens: sometimes these are in the form of a "furniture system." Computer terminals are dotted about on desks, often with printers near them; there are copiers against the walls, telephones on the desks, and wires and cables wind their way around table legs and across the floor into raised "tombstone" outlets. There is tired dry air and pale carpeting of an indeterminate color that shows signs of wear. There are often large windows, the glare and heat from which are prevented by drapes or louvers, so that often there is no view. There are often overlit windowless conference rooms and boardrooms that stop being ventilated properly once the room is in use and the doors are closed. Sometimes there are a few plants.

1-2. A typical open-plan office interior.

There are also fans, heaters, ionizers, and smoke-eating machines, usually brought in by office workers who want to feel better as they do their work. In some offices such additions are prohibited by building managers or by the organization. And in others, instead of bringing in their own "environmental aids," workers complain about physical discomfort and symptoms of ill-health. Regardless of the way an office building looks from the outside, for those who occupy it, the modern office can be an anonymous, uncomfortable, discouraging, and sometimes unsafe environment. Many office workers will confess that if they get their work done there, it is *in spite of* their office space, not because of it. From a manager's point of view, workers who have to deal with adverse environmental conditions consume energy that could otherwise be spent on their work or could benefit the social environment of the office. The modern office, for one reason or another, seems to be a long way from actually encouraging its inhabitants to do their work.

In some buildings, adverse environmental conditions are even more serious. A few years ago, one of the largest and newest office buildings in Canada—the Terrasses de la Chaudiere near Ottawa (fig. 1-3), a huge, elegantly designed, red-brick, multistory office building completed in 1979—was

1-3. The Terrasses de la Chaudiere.

picketed by office workers, who marched around the front entrance with banners. They refused to continue working in the building on the grounds that the interior air was polluted. They maintained that people were getting sick from the building, and they demanded that some action be taken to improve the quality of their work environment. The first complaint about worker illness from polluted air inside the building became public in fall 1979. Preliminary testing of indoor air samples by a consultant team of experts found that conditions in this large new government office building were acceptable, but workers' complaints intensified. By November people were taking lengthy sick leaves from work, and some worked at home and did not come in at all (*The Citizen* 1979).

In spring 1980 a new wave of complaints was heard about the "stench" pervading the building. This time an investigation discovered a missing drainage trap at the bottom of one of the elevator shafts. This construction oversight apparently caused nausea, dizziness, and headaches among the office workers (*The Citizen* 1980, 3). Their symptoms, and their complaints about bad air, persisted long after the error was corrected. Workers told the press, "If I don't get out of here for an hour a day, I'm just wiped out"; and "The air just chokes me up by the end of the day." About a year later workers' persistent complaints resulted in more testing of the indoor air, and dangerously high levels of formaldehyde and trichloroethylene were found. By this time the building was fully occupied, and some 6,000 workers were affected. There were reports of uncomfortable indoor temperatures, lack of air, noisy ducts, headaches, exhaustion, and sore throats. The cause for this potentially toxic pollution was thought to be Urea Formaldehyde Foam Insulation (UFFI), but its installation during construction was hotly denied by the building developer (*The Citizen* 1981, 2).

By the fall of 1981 newspapers were reporting workers' claims of higher rates of miscarriage and risks to pregnancy in the Terrasses de la Chaudiere than for comparable groups of women. In 1982 a woman named Vera Wall was finally awarded compensation for having to take thirteen days of sick leave the year before because of "respiratory problems, throat and ear infections, headaches, nausea, and severe fatigue," which she claimed were the result of interior building conditions (*The Citizen* 1982). This paved the way for several other compensation claims.

Workers in the building were still complaining about their office environment in 1985, seven years after the first complaints were publicized, and some eight to ten years after the building was built (*The Citizen* 1985). One of the last studies commissioned to analyze possible causes of workers' ill-health did some detective work in the bowels of the building where the air handling system mixes the air drawn in from outdoors with a portion of the warmed air returning from circulating through the building. It found that the mixed air chamber had not been properly built: large columns and chunks of concrete effectively blocked the flow of renewed air back into the building, so the air being supplied into the office space was not being refreshed by the new air from outdoors and was not being supplied in the proper volume. The study concluded that the problems reported in the building "are entirely typical of the many other episodes of 'building illness' or 'tight building syndrome' investigated so far without satisfactory explanation in North America and Western Europe during the last decade" (*The Citizen* 1984).

In addition to whole buildings with long unhappy histories like the Terrasses de la Chaudiere, specific building conditions generate employee discomfort and complaints in offices all over the world. In Melbourne, Australia, for example, a group of government workers refused to go back to work at their video-display terminals (VDTs). They reported suffering from bad lighting, uncomfortable furniture, too many hours immobilized at their desks, and symptoms of Repetitive Strain Injury (RSI)—muscle pain and joint swelling—in their wrists, arms, and necks. Their union took up the case, demanding regular rest periods, illness compensation, and environmental improvement. Eventually, their work area was redesigned, with shaded lights and ergonomically designed work stations.

In another example, a small group of office workers on one floor of a multistory office building were regularly forced from their work area once or twice a week because of nausea and faintness. These symptoms turned out to be caused by fumes from a kitchen exhaust that emanated from a shaft in the interior of their building. In the area where they worked, there was a small access door to the shaft through which the fumes seeped. Their union was not as strong as the one in Australia, and nothing much was done. The workers continued to live and work with the problem.

In another modern office building a woman left work early repeatedly with headaches and eyestrain that prevented her from working on her clerical task at her desk. Various attempts were made to solve the problem, including the purchase of eyeglasses, but she remained unwell and unproductive. Eventually, the plate glass covering her desk top was removed. Her headaches and vision problems improved, and she was able to perform her work normally. The glare caused by bright overhead lighting reflecting off her work surface had rendered the close visual task all but impossible for her to perform.

Offices or housing built on top of shopping centers have problems with noise and vibration and with carbon monoxide and petroleum fumes. Whole office buildings have been emptied out because of faulty air handling, indoor air pollution, or fear of building-transmitted diseases, such as Legionnaire's Disease.

Even if these conditions are rectified, and repairs and renovations remove sources of discomfort and ill-health, the reputation of the building among office workers is not so easily repaired. Once the people in a building have experienced problems of this nature, their sense of security about the building is not the same. They are suspicious. Even if the walls are painted in bright new colors, lighting is improved, noise is reduced by introducing a sound-masking system, and plenty of good clean air is provided, people do not easily renew their faith in the quality of their work environment. The occupants of the Terrasses de la Chaudiere are still complaining. Their experience has been that solving one building problem gives way to the presence of another. The question is, can such serious building problems as those in the Terrasses ever be solved? Office workers and their managers, not just in troubled buildings but in all office buildings today, find themselves wondering just how safe (meaning healthy) their building environment is. Concerns about sore throats, coughs, eye irritation, and fatigue at work take on new, menacing overtones when considered in the context of indoor air pollution, improper lighting, and VDT radiation in modern offices. Once office workers become aware of these threats to their comfort and health, they are unlikely to feel comfortable and healthy, no matter how many times they are assured that the air, heat, light, and noise levels have been tested and/or improved, and that there is no risk. For those involved in the daily decision making and management of office buildings, the problem of lost confidence among users can lead to an increase in their workload and in the difficulty of their job. Building users increasingly complain and demand building improvements. Worker morale is low and productivity can be reduced.

If poor-quality office buildings cause anxiety and low morale among workers, create extra work and worry for building managers, are costly to organizations in these and other ways, and have long-term effects on how the public views office buildings, how do office-building problems come about and why are they not being corrected? In urban areas large corporations often pay for stylish, glamorous, and image-making buildings for their headquarters, and developers spend more of their construction budgets creating elegant-looking buildings than creating comfortable, supportive, well-planned interiors. But as the office-space market becomes more competitive, as is now happening in many North American cities where the market for office space is almost saturated, even small-scale developers, who build most of the office space in the United States, have to address issues not just of building looks and style, but of basic building quality (Birch 1986). People should be able to work productively and with some enjoyment in buildings that pose no threat to worker health and well-being, and soon they will demand to do so.

HOW DO BAD BUILDINGS GET THAT WAY?

Why are mistakes made that generate months of discomfort for building users, take weeks to repair, and years to understand? There are many ways to answer this question, but perhaps the most useful one is to examine the way buildings get built in our society. This examination reveals how the responsibility for quality control shifts from one professional to another during the process of creating a building. There are many opportunities for slip-ups, oversights, carelessness, and irresponsibility in this process as it is typically carried out today. How is quality controlled in modern building construction, if it is controlled at all? One view of quality control (or the lack of it) in the building industry is described below.

The individual who wants to develop an office building for his own use or for rent or sale must first assemble and acquire the land. The costs of building and then leasing or selling, or of moving into and operating the office building constructed on that land are added to the cost of the purchase to provide a total estimated dollar outlay against which the owners or developers must calculate their return. The amount of profit they make depends on how much habitable space they can create in the building or buildings they put on the land, and how quickly they can get it or them occupied, rented, or sold. Thus, the developer generally selects a building form that provides the maximum habitable space relative to the overall amount of space built (gross-to-net ratio), using local zoning provisions, current technology, and available materials. An architect may be hired to design the building and to supervise its construction.

During this initial decision-making period, numerous trade-off decisions are made about the basic form and style of the building, about how much of the budget will be spent on "extras" in the form of ornament and decoration, and how much on amenities such as parking or elevators that go beyond basic code requirements. Environmental elements must also be decided upon, such as whether or not to incorporate an atrium in the building design. An atrium is one of those design features that has the important function of determining the style and image of a building, and thereby a large part of its marketability, but cannot be rented out or sold as functional, usable space. How much front-end investment is such space worth? The capital outlay and decisions on how to spend front-end money should really be made along with anticipated operating costs in what is called *life-cycle costing* of the building (Bon 1983).

However, economic circumstances often dictate that the present is traded off against the future, and short-range decisions that reduce front-end costs dominate the design and construction process. Developers who build office buildings speculatively, or quasi-speculatively, may not have a long-term interest in the functioning quality of the building for users, so their cost decisions are likely to forego long-term quality altogether in favor of short-term good looks (Brandon 1984).

Recent revelations from the Canadian government indicate that a local developer who built several major government office buildings balanced his projected budget for each building according to how much the government had agreed to pay him to rent the space once the building was built (*The Citizen* 1985). In effect, the government signed a contract that enabled him to know exactly when and how much his return on investment would be, and over what time period. This is not unusual in large urban real estate development, where banks lending money to developers for a project are often major tenants of the built complex (Lorimer 1978).

In this case, the developer knew not just his *projected* return but his *actual* return on investment before construction started. His client (the federal government) did not elicit building-quality promises in return for the guaranteed rent; very few clients do. Because the deal was struck in a noncompetitive environment (the government did not offer the project to other developers), this developer had little incentive to control building quality. In fact, he had every incentive to make cost-cutting decisions during design and construction. The buildings he built for the government have turned out to be problematic and expensive to maintain and operate; one of them is the Terrasses de la Chaudiere. But the costs of repair and renovation are not paid by the developer. The dollar costs of correction and repair are hidden in the buildings' operating budgets and are borne, in the case of government buildings, by the taxpayers. The health and comfort costs are borne by the workers whose offices are in this building and the others like it.

Some environmental difficulties and disasters of modern office buildings can be traced to poor design decisions on the part of the professionals involved and/or bad cost judgments on the part of their clients. During the design and construction phases of an average building, some developers will stay involved with their projects to monitor cost and quality decisions, but many assign what might be termed "quality control" decisions to the architects, preferring themselves to supervise cost expenditures. The architect, then, is often at the mercy of his own consultants: the engineers. He does not understand their technology or the computer programs that provide them with their design specifications; he accepts the product they supply for the building as a state-of-the-art design decision.

In the case of mechanical ventilating (HVAC) systems, engineers apply routine formulae to the design of air distribution systems throughout a building. The same designs are increased for large buildings and reduced for small buildings. The amount of square feet to be ventilated determines the length and width of the ducts, which in turn determine the power of the fans. The more standardized the specifications, the cheaper the system is to design and install. The less known about the specific requirements and ultimate use of the space, the more standardized the HVAC specifications can be, and the less appropriate to the ultimate uses of the building they often turn out to be.

The formulae electrical engineers use to supply lighting are similar: uniform amounts of light are supplied throughout the building to meet "worst case" (i.e., no daylight) conditions, such as night. Variables such as the fact that most people do not work at night, that their office tasks vary, that their eyesight differs, and that different interior materials and furnishings will be used, are not incorporated at this stage of design because they are not known, and even if they were they would generate unstandardized requirements, thus increasing the cost of the lighting and electrical system installation.

Architectural and engineering decisions are often made with little reference to each other, so an attractive architectural space such as a high, glassed-in, sunny atrium may have no mechanical means to exhaust the heated air that collects at the top. Even the engineering decisions may fail to be carried through to construction, because the builder has his own agenda and priorities to get the job done: the larger and more complex the building, the more last-minute changes the builder makes to accommodate unforeseen conditions that could add to the construction time and costs of the building.

Once the building starts to be built, the contractor is in charge. If there is a responsibility for quality control at this stage, it shifts by default to him. As the building is constructed, numerous changes are made, some of them in the form of change orders approved by the architect and the client, but some simply slipped in after some verbal discussion. Design specifications can, by this procedure, be subtly and not-so-subtly altered. If the chunks of concrete left in the mixed air chamber cost too much to remove, they stay. If the metal parts for the ventilation ductwork cannot be purchased in time or are too expensive, perhaps plastic parts are substituted. If only two drainage traps are delivered instead of three, one is left off, since to wait for it would cost time and therefore money. The ultimate costs of decisions like these fall on the eventual occupants of the building, whose employees have indoor air-quality problems and who have to pay for the solution.

The modern practice of "design-build" construction projects, in which one company designs the building and then executes construction to reduce the number and variety of agents involved in the process, can also lead to critical building decisions being made in a disjointed and incremental fashion. It was a lack of monitoring of the overall process that, some think, resulted in the building failure that caused the collapse of the partially built L'Ambiance Plaza in Bridgeport, Connecticut in 1987, killing twenty-eight construction workers (*Boston Globe* 1987).

Once construction is complete, office-space planners and interior designers enter the completed building to prepare the office space for use. Working with the building owner or the tenants who will be leasing, they make design decisions based on how the space will actually be used. They erect walls and partitions, design colors and supply furniture, and locate heat- and fume-generating equipment and machines. In most cases they have no formal contact with the developer, the architect, the consultants, or the builder. So it is that printing equipment or microfilm machines are installed without any special ventilation, causing workers to complain later of heat and odors. Or a blueprint machine (which exudes ammonia fumes) is provided with its own exhaust that somehow does not vent directly to the outdoors as it should, but directly into the ceiling plenum, the space through which air is carried to the mixing chamber to be recycled through the building. This kind of error places whole building populations at risk from small amounts of toxic fumes. Perhaps an office is walled in with an air supply vent in the ceiling, but no air exhaust, so that whenever the door is closed for meetings, or even for private use, the room rapidly becomes overly warm and stuffy. Users therefore have to hold their confidential meetings with the door wide open. These kinds of problems multiply as the building occupants move in, settle down, and make their own moves, changes, and relocation of workers and equipment.

Another quality control opportunity arises in the new building after the users and their equipment have moved in. The engineers are supposed to return to the building to balance the air handling systems and to make sure that the amount and temperature of air supplied conforms to the actual use of the space in terms of numbers of people and types of equipment, and how they are distributed in the building. However, this balancing procedure is often overlooked. If it is performed at all, it is often superficial. Sometimes it is performed after the construction crew has left the building and before the people have actually moved in, which defeats its purpose. By the time people move into the building, many of the building problems that will plague them are already "built in" (Levin 1986).

QUALITY CONTROL FOR BUILDING OCCUPANTS

After people move into the building and start to use it, quality control in the environment falls to them, regardless of whether they own or lease the space. It is not surprising that most building managers and facilities staff complain that all their time and budget is expended on maintenance and repair, on *reacting* to building performance problems of one kind or another.

The first year of occupancy of a new building is often costly and difficult, as many small (and not-so-small) building mistakes are repaired, nonfunctional equipment and systems are made to work, and near disasters are averted. The responsibility for this stage falls heavily on building managers and their staff, who often have neither the budget nor the personnel to carry out the work that is needed to make the building habitable. The most glaring and blatant of the building's problems are taken care of first; after that the accommodations staff have to fight to be heard by senior management for money and resources to bring the building up to an acceptable level of environmental quality. By that time, however, workers are often resentful of and frustrated at the adverse building conditions in which they are supposed to be working to the best of their ability. Problems of design and construction can be exacerbated by other conditions that are the result of ignorance or oversight by both users and building managers. Every time a major interior move is executed, building managers have to solve new problems of walls and partitions inappropriately placed and preventing air flow, of the wrong lighting for a task, of new acoustic problems, and of ventilating/heating equipment.

Why do the occupants of the building end up bearing the brunt of the oversights and errors of the development, design, and construction professionals? With most professional services, the consumer who is dissatisfied with what he receives has recourse either to other members of the profession, to whom he may take his business, or, if there is a question of incompetence or negligence involved, to a court, and, ultimately, to peer review of the individual's or company's performance. In the case of an office, the building is there and in place, occupied and used. The consumer did not (for the most part) commission the building and therefore is not in a position to take his or her business elsewhere. Many building problems do not get as far as a courtroom because—unlike Vera Wall of the Terrasses de la Chaudiere—the users are not in a position to determine when and how they have been wronged. Most people simply do not believe that the typical environmental problems they experience every day at work do not need to be there, that office buildings could and should be better, that the technology to meet higher standards of building quality is available but that the consumer demand is not. In other words, most building occupants believe that nothing truly substantial can be done to improve the office in which they work.

Problems in office buildings are not so much the result of incompetence and negligence as they are the "fallout" from an incoherent and unmonitored process where each decision maker trades off the commitment to a high-quality product for the chance to make a buck. A recent international comparison of office building quality between the United States, Britain, and West Germany concluded that "We are not achieving great quality [because of] conflicts of interest between the different agents of the production process" (Ellis 1986, 7).

The evidence of incompetence or negligence in a poor-quality building is difficult to prove. The office building problems described earlier in this chapter are typical and result in substandard work environments for people who work in offices. Without the demand from consumers for better-quality environments, the building and development industry is not motivated to improve its product. And as long as the building is not falling down (like L'Ambiance Plaza in Bridgeport), or losing its windows (like the John Hancock Building in Boston), or actively poisoning people (like the Terrasses de la Chaudiere), how are building users to know that they do *not* have the best possible product?

Quality control in the building industry compares unfavorably to quality control in almost any other industry. Construction of large engineered products, like airplanes and tanks, is monitored in detail, with all parts tested and retested, both independently and as a component of a larger system. Money is spent on a large scale by these industries on R&D, where there is a major thrust to create technology that will alter both the process of creation and the ultimate physical form and appearance of a large, complex, manufactured artifact. Not so in the building industry, where large-scale R&D is more like a burp than a thrust.

One so-called innovation is the "intelligent building," in which feedback from interior conditions directs computerized correction and adjustment of the systems. Users can telephone for changes in temperature, or to switch lights on and off, and emergency systems communicate directly from the alarm to various levels of building response (*Facilities Planning News* 1986). This does not offer a fundamental technological breakthrough so much as a computerization of existing building technology. An office building's systems can be computerized much as libraries have become computerized. Libraries store books and other reading material according to a system that enables each item to be identified and retrieved. The fact that the identification for access and retrieval is computerized does not alter the physical forms of book storage nor the basic ways in which people use libraries. In modern office buildings, there are likely to be limitations on environmental quality whether the building is "intelligent" or not.

THE SOCIAL MEANING OF OFFICE BUILDINGS

In our society buildings are value statements vis-à-vis the occupants of the buildings (Foucault 1977; Lindheim 1974; Vischer 1975). For example, the architecture of mental hospitals makes strong statements about the way society views mental illness and the mentally ill. Schools, prisons, and even housing have changed in form and appearance over the centuries in accordance with society's changing values regarding school-age children, convicts, and families. The modern office building implies definite assumptions about what we as a society think of the white-collar work force; that is to say, the mostly clerical women and mostly male executives who put in an eight-hour day doing paperwork, speaking on the telephone, and sitting in meetings. At the very least, the modern office building carries statements regarding two sets of values, perhaps implying two sets of users.

On the one hand, office buildings can be elegant and

attractive. They are aesthetic statements by leading architects. They have image-making features such as spacious, daylit atria, waterfalls, tile and glass entrances, soft carpets, Muzak in the elevators, hangings and tapestries, designer colors, and above-grade enclosed pedestrian connections to other downtown buildings. To a casual visitor, or a passer by, such buildings represent the wealth and power of our corporations and our government, the success of free enterprise, and the sweetness of being a winner.

On the other hand, for the workers in these glamorous creations, this experience does not always carry through to the daily routine of working in the building. The job, to be sure, is clean, above ground, and is not manual labor. It is often well paid, it is respectable, though often anonymous, and it is mostly secure. But there are unknown risks to health in the sealed building that recycles pollutants into the air supply: headaches from the bright lights; a feeling of fatigue in the afternoons; an itchy dry throat and sore eyes from the dry dusty air; and irritation with the slow, crowded elevators, the constant noise in the office, the inadequate privacy for working alone and for conversation, and the lack of parking. Small improvements in technology and sensitive interior planning could do much to improve the quality of these office interiors, but if money is spent on these buildings at all, it tends to be spent on the façade, the structure, the look, and the architecture, all of which contribute to the building's public image, and none of which has a very direct impact on the office-workers' quality of life.

One obvious conclusion is that people who work in offices, especially clerical and support staff who do routine jobs, often in deep-floor open-plan layouts, are not very highly regarded members of society. They are dispensable; they are replaceable. However, if adverse office conditions affect highly placed professional workers, executives, and management, the whole organization is seen to suffer. Many organizations encase their executives in enclosed offices with windows to ensure, in fact, that they are as little affected as possible by the problems of the building environment. But this is not really a solution. Organizational health and success depends on the satisfaction and productivity of *all* work groups.

OFFICE-WORKER PRODUCTIVITY

In what ways does the quality of people's work become measurably better as a result of an improved building, managers might ask. If morale is low, or people are leaving early from work, or taking too many sick days off, organizational productivity may be affected by a bad building. The extent and nature of this effect has been much debated in the literature. How much money is it worth investing to make an office into the optimum work environment for its occupants?

One way of making decisions is to construct a "cost-benefit" framework for analysis of the costs and the benefits of alternative solutions to a building problem. For example, if people are complaining about poor ventilation, a series of decisions must be made by building managers. First, are the complaining workers right or wrong? Then, if they are right, how bad is the poor ventilation: is the air carrying toxic pollutants? Or is there an inadequate number of air changes per hour and a build-up of carbon dioxide? Or is there dust in the air or dryness irritating people's throats? Third, what is to be done about it? Major changes to the air handling systems are likely to be expensive. Moving people to another part of the building may be inconvenient and may not solve their problem. Adding fans or opening windows may help workers' morale but is not likely, by itself, to improve their health, if this has been adversely affected by the poor ventilation.

The solution selected is likely to reflect the degree of seriousness of the workers' complaints. If they are "dropping like flies" and the incidence of sick leave is taking a dramatic rise, an expensive retrofit of the mechanical systems is not unlikely, and the dollar cost is worth the benefit of reducing sick leave and avoiding legal action. If the workers continue, in spite of their complaints, to work reasonably productively, cheaper solutions (or none at all) may be selected. The difficulty facing the organization is knowing when (and if) the performance of workers is affected by bad environmental conditions. In the majority of modern office buildings, environmental conditions are not so poor as to cause obvious ill-health; on the other hand, people often complain of general discomfort and a feeling of dislike for their office environment. Managers may wonder if such feelings reduce workers' ability to perform during the workday.

In formulating the cost-benefit analysis of solutions to building problems, the cost side of the equation is usually estimated in terms of capital to be expended: the dollar cost of renovating, reconstructing, or otherwise changing a building. The benefit part of the equation is usually expressed in the less tangible terms of office-worker performance or job productivity: their value to the organization is computed in terms of how much work they yield for the dollars paid to them, or how much money the company makes from the work for which an employee is paid.

Another type of formula is the "Productivity Indexing" used by Northern Telecom, which is designed to measure quality as well as quantity of work output. In this system, each work group generates measurable indicators of its goals and objectives and scores them over a three-month period to calculate a single "Productivity Index" for that time period. Measures of white-collar productivity often apply not just to the individual worker but to the work group and to the organization. To the extent that individual workers rely on each other to get the job done, a more useful concept is that of *organizational effectiveness*. Organizational effectiveness has been defined as the result of individual and group productivity, of individual job satisfaction, and of an overall "high quality of working life," which includes the spatial and sensory environments of the office setting (Ward et al. 1984). The criteria for assessing increases in organizational effectiveness resulting from building improvements are even less amenable to dollar formulations than those for individual worker productivity: how is a manager to decide how much to invest in environmental improvement to ensure a "payoff" in organizational effectiveness and individual productivity?

Current reports indicate that the North American work economy is based increasingly on the provision of information and services. Whereas in 1950 white- and blue-collar workers each constituted a 30 percent share of the labor force, the white-collar figure presently stands at 53 percent of the work force of the United States, representing a 70 percent share of the total U.S. payroll.* The white-collar work force is expected to grow to 65 percent of the work force by 1990 and to 80 percent of the total U.S. wage-earner group by the year 2000. In 1982 the breakdown of white-collar workers was as follows: management and administration—22 percent; clerical—35 percent; and sales—12 percent (American Productivity Center 1982). New electronic technology and computerized communications are causing rapid changes in the nature of white-collar work. As new information technology changes the organization of work in office buildings, office-worker productivity becomes more difficult to quantify. Traditional productivity measures derive from an earlier era, when the industrial engineering analyses of manufacturing processes measured tangible output by workers.

While the measurement of manufacturing productivity has developed its own techniques, which are widely used and well respected (e.g., time and motion studies), few of these techniques are directly transferable to the intangibles of conventional office tasks. There is little agreement in the business world on exactly what office-worker productivity means. One recent study suggested that traditional concerns with office-worker productivity are being replaced by the need to accommodate computers and the effects of computerization on office work (Steelcase Inc. 1987). To move from the hard-to-define area of white-collar productivity to an understanding of how important the physical environment of work is to people's productivity is even more of a challenge. However, recent studies of the work environment and its effect on satisfaction, on worker effectiveness, and on individual performance, some of which are described below, have risen to this challenge.

A 1982 survey of 140 American firms found critical relationships between three segments of the office environment: human resources (the people), automation techniques (the systems, the tools), and environmental design (the habitat, the place, the work environment) (American Productivity Center 1982). The study defined the office as any place where productivity is focused and found that optimum productivity depends on the

---

*In 1982, U.S. payroll totalled about 1 trillion dollars.

integration and synergy of all three segments. The survey revealed that there was little solid information on worker productivity in white-collar organizations, and even less on how it should be measured. Evaluations of specific programs to enhance productivity showed that a 16 percent improvement in efficiency and effectiveness could be attributed to automated office systems, a 12 percent improvement to "human resource development," and another 12 percent improvement to better environmental design. The authors concluded that, "The opportunity for financial improvement within the white-collar sector is enormous. Given an annual cost of $1 trillion per year, American industry would save $95 billion if they put productivity improvement programs to work." Using the study's figures, improved office design could be responsible for nearly $32 billion.

Another report presents a three-site study of three different organizations with "before" and "after" conditions of building improvement (Steelcase Inc. 1983). Two of the studies include measurements of employee productivity. The authors found that

> An average productivity increment of 3.92 percent was determined for individuals at the Boston office building, while group rates went up by as much as 17 percent. In the Tech Center study, overall employee performance ratings went up a full 5 percent, compared to an increase of 1.6 percent in a control group. Data from the IRS is even more impressive: using Methods Time Measurement, a productivity increase of 7.4 percent was calculated. Actual work-counts at this office, however, have put the rate of improvement at approximately 9.4 percent.

The study attributes these productivity improvements to "environmental changes," although it is difficult for the reader to distinguish between *physical* environmental and *social* environmental change.

A 1983 Rand Corporation study tried to identify ways in which information technology was successfully incorporated into white-collar office work in twenty-six different organizations (Bikson and Gutek 1983). In analyzing companies where innovative information systems had been installed for at least six months, the study concluded that characteristics of the building environment have "considerable influence" on the implementation of office technology and can significantly affect the success of organizational innovation.

Industrial psychologists and efficiency experts have traditionally concerned themselves with issues such as job performance and job satisfaction, task analysis, and management processes. One of the first studies to link job performance and job satisfaction with the quality of the physical environment was carried out in the late 1970s for Steelcase, a furniture manufacturing company (Harris and Assoc. 1978, 1980). This research linked worker comfort and worker performance by questioning office occupants on their degree of comfort with various environmental conditions (e.g., temperature, lighting, and seating); it concluded that people would do more work in an average workday if they were more physically comfortable. Those workers

> who more often associate improvements in comfort with productivity are employees who work in a bullpen (pool) type of office (63 percent), regular workers (58 percent), persons employed in the business and professional services industries (60 percent), and persons employed in the banking and investment industry (61 percent) (Harris and Associates 1980, 67).

One of the most extensive studies by design professionals of the relationship between the office environment and worker productivity was carried out for Westinghouse by BOSTI, a Buffalo-based environmental research firm. (BOSTI 1982). The findings from this five-year analysis of questionnaires from more than 4,000 office workers and of field measurements of environmental conditions demonstrate that office environmental design affects both job satisfaction and job performance. The analysis identified the dollar value of productivity gains from selectively improved workspace conditions for managers and for professional/technical workers at about 5 percent of annual salary. Thus, the dollar gains from specific environmental improvements could actually be stacked on the benefit side of the cost-benefit framework. This is one of the few studies to attach quantifiable dollar benefits to environmental improvements, although it has, in fact, been criticized for overemphasizing environmental factors at the expense of social and organizational influences on worker behavior (Ellis 1985).

Preliminary findings from these studies and others like them indicate that the physical environment for office work may account for variation of 5 to 15 percent in employee productivity. On the basis of a $1 trillion white-collar annual salary base in 1986, and using the most conservative estimates of the above studies, a 5 percent productivity increase (or cost reduction) arising from improved design of the office work environment would amount to $50 billion per year in the United States alone. The long-term yearly functional use costs of an office building are principally the costs to an organization of its employees' salaries. These costs exceed both the yearly, long-term amortized construction costs and the yearly plant operation and maintenance costs by a multiple of five to ten or more. Therefore, from the standpoint of total long-term productivity of the office building, each 1 percent increase in the productivity of the office worker due to an improved physical work environment is as effective as a 5 to 10 percent increase in total building "productivity" or effectiveness resulting from a cost reduction in building materials, systems, construction, maintenance, or operating costs. Put the other way around, the investment of one dollar in building improvement could, according to the results of these studies, result in ten dollars worth of improvement in the performance of the organization. Of course, not just any old improvement will do; the careful planning and defining of which aspects of the environment to improve is an implicit condition for the success of this argument.

SELECTING BUILDING IMPROVEMENTS

The effect of the building environment on worker morale and productivity is being debated more these days now that examples exist of office buildings driving out their workers altogether. White-collar worker's unions are introducing "quality of life" clauses into contract negotiations. Occupational health researchers are becoming more involved in the prolonged effects of adverse building conditions on worker health.

The importance of physical improvements to office buildings and the work environments they create cannot be ignored. At the very least, productivity should not be *reduced* by ill-health or discomfort caused by building conditions; at best, productivity can be increased in most modern offices by environmental improvement. The question is, what kinds of improvement should be made, and who should make these decisions? Given what we have learned about the problems people have in office buildings and the way in which these problems come about, what should be done to improve the environmental conditions in which people work? How can an average state-of-the-art modern multistory building be made to function as a positive and noninvasive context for the optimal performance of complex office tasks by large numbers of people?

Office building owners and managers face this question each time they must set priorities on spending their maintenance and operating dollars. Is a complete and expensive rebalancing of the ventilation system a better way of allocating resources than changing the light fixtures, replacing the carpeting, or relocating all the VDTs? Does the fact that people on one floor complain about indoor air quality mean that more (expensive) air sampling testing should be done, or that in fact this is a work group that dislikes its supervisor, or that cannot make telephone calls without being overheard, or that lost their parking space privileges a month ago?

How does a manager go about ascertaining the cause of users' complaints and, by extension, his building problem? Once determined, how should the information be used? Will specific tests of building performance yield a better solution than a focused group interview or a questionnaire survey? And when does the cycle of problem-testing-correction stop? Can a building's environmental design ever be fully responsive to workers with so many different possible job requirements? If it cannot (which appears to be a realistic assumption), then how can a building manager tell his superiors that all this money is being spent on continual improvements to the building? Will it then one day be perfect? And in fact, how are managers to know when to stop testing, measuring, repairing, and improving aspects of the building because things can get no better? Some buildings have been built so badly that no amount of money will make them truly good environments. Others can be brought to a much improved level of quality, but unless one wants to throw buckets of money at the building to do so, careful study is necessary to ascertain on which problems the improvement dollars will best be spent. In still other office buildings, of which the Terrasses de la Chaudiere might be an example, the users are so jaded that nothing less than a different building will satisfy their standards of environmental quality.

One question that is basic to this debate is whether or not there is a *measurable quantifiable* impact of better office design on people. From a decision maker's viewpoint, if money is spent on improving a building to make people more satisfied, then the quality of their work should be measurably better. Conversely, a poor building is one that does *not* provide an appropriate context for the work people are doing in it, no matter how much money and time have been spent designing it. A good-quality environment occurs when the building is properly supportive of and suitable to the tasks and objectives of the workers. To answer the questions posed by building managers, therefore, it is important not to consider worker behavior independent of the building and building performance separate from the users: both must be considered together.

Most people in the building business would find it helpful to be able to measure or somehow systematically assess quality in buildings. If we do not understand environmental quality, we cannot discuss the possibility of environmental improvement, because what are we making improvements to when we repair and change an environment if not its quality? Improving something means increasing its quality. We may not all agree on what quality is, but we all know what we mean by an improvement: something is better than it was before. In the case of improving the office environment, that which is better is the *interaction* between the people, their work, and the space in which they do it. Simply making repairs, as was done in the Terrasses de la Chaudiere, does little to improve workers' perceptions of their lot. For building problems to be solved properly, the way people use and feel about the building is part of the definition of the problem and needs to be taken into account. In describing, for example, a pair of pliers to a visitor from a long-lost Stone Age tribe, the information that they have two "arms" and can open and close on a nail or a screw is of little use without the additional information that they fit into a person's hand and are worked by the person's fingers. Similarly, what does and does not contribute to productivity in an office environment cannot be reasonably considered outside the context of how the workers are using the building.

This approach to the understanding of buildings and of environmental quality in buildings is known as the *building-in-use.* The building-in-use is the entity with which building managers are involved: not just the building, its structural and mechanical systems, its shape, size, finishes, and materials, its heat and light, its furniture; but an occupied, operating, functional building that is brought to life by the people working in it and the tasks they carry out there.

SUMMARY

The majority of the North American labor force works in offices, and this percentage is on the increase. Many modern office environments are, if not actual threats to health, uncomfortable and less than optimally desirable places to work. Workers often complain about environmental conditions in offices, either locally to their managers, or publicly to the press. Most of the knowledge and technology to improve environmental design in office buildings exists in the building industry, but for reasons having to do with the structure of the industry, the nature of the building commissioning process, and the ignorance of building users, much of it is not applied. Many opportunities for quality control during building design and construction are lost.

Adverse environmental conditions in offices have costly direct and indirect effects on organizations. These include low worker morale and reduced employee performance and organizational productivity. Studies indicate that improvements to the work environment could result in substantial dollar increases in organizational effectiveness. However, senior corporate decision makers often require "proof" that dollars invested in building repairs and renovation show tangible evidence of better worker performance. In many cases, improvement in a state-of-the-art office environment could be achieved with relatively little expense and a broader understanding of the psychology of people working in such buildings, resulting in more satisfaction all around (lower operating expenses, happier workers). The purpose of this book is to show how this might be done.

The building-in-use is a system of which the building itself and building users are integral components, neither of which can be fully understood if considered separate from the other. How this helps define environmental quality, the meaning of environmental improvement, and a better understanding of the building-user relationship are the topics of the next chapter.

Chapter 2

DEFINING ENVIRONMENTAL QUALITY: INTRODUCTION TO THE BUILDING-IN-USE

In the last chapter we discussed the difficulty of deciding how to make a building better without having a definition of what a good building is. In this chapter we probe the meaning of environmental quality and discuss how it could be improved in offices. Our definition of environmental quality is predicated on the concept of the building-in-use, which is to say, the concept of a building as a changing system of interdependent physical features and human activities. This makes quality complex—but not impossible—to measure, as it is a dynamic rather than a static concept. This dynamic definition makes static measures, such as the application of standards and analytic techniques of measuring building performance, into incomplete indicators of environmental quality in buildings.

USERS' COMPLAINTS AND BUILDING PERFORMANCE

People who work inside office buildings day in and day out often feel they would benefit from better lighting, less noise, fresher air, or some other improvement in environmental conditions, such as more comfortable furniture. In most cases, these workers are not able to open their windows, turn off their lights, close their doors, move their desks, or turn down the heat. Their concerns about these and other restrictions on their behavior are expressed through complaints to building managers and operators.

For those concerned with managing a large and complex office building, the principal gauge or tally of poor building performance is often the number and severity of complaints they receive from building occupants. To improve building operation, those responsible for the building (including owners) must make decisions about when and how much change to make to the building to create a more satisfactory and comfortable work environment in response to users' complaints. Often they do not know how the effects of the operating decisions they make affect the workers in the offices.

For example, building managers who decide to save energy by removing bulbs from overhead fluorescent light fixtures may find their staff overwhelmed with requests from workers for desk lamps. In one building, a building manager who installed a "white noise" background sound-masking system to improve voice privacy in open offices was obliged to switch it off because he received so many complaints about excessive ventilation and drafts from workers who confused the low hiss of "white noise" with the background noise typical of mechanical air handling systems.

Relying on users' complaints is an ambiguous method of assessing the scale and importance of a building performance problem. First, it means that a building that is good and works well has no equivalent way of being assessed or even noticed. Who sends in positive comments about good-quality office buildings? Second, when workers complain, it is difficult to determine what they are complaining about. It could be their supervisor, or their insomnia, or their neighbor, as well as any building problem. Moreover, people who are predisposed to complain in the first place may complain equally about their dissatisfaction with a color they dislike or the distance of their desk from the window as they might about headaches and sore eyes, or faintness and nausea from an indoor air pollutant that might cause cancer. To use people's comments and complaints as an indicator of their reduced productivity is unreliable, at best.

Third, building managers' solutions to office environment problems as defined by users' complaints do not necessarily result in improved worker morale or productivity. In fact, where building problems are of long standing, low morale among workers can persist even when the problem is corrected. People simply find something else in the building to complain about. When this happens, the faith of senior management in their building management staff decreases; building managers drop even lower on the organizational totem pole and are listened to even less on the next occasion.

THE DILEMMA OF BUILDING MANAGEMENT DECISIONS

Workers' concerns about their building rarely rise to the senior managerial ranks of an organization, unless there is a real threat to worker health. They *do* fall unceasingly on the ears of the people who manage buildings. For managers, the problems of maintaining and operating good-quality buildings are myriad and neverending. Although money is spent by owner and tenant organizations to fix up office buildings that have been badly designed or constructed, it is difficult for building managers and organizational decision makers to know how much they should invest in the office environment. Changes in individual productivity are hard to measure and even harder to prove as resulting from environmental design improvements. Building management responsibilities and resources are often confined to "putting out fires" and fixing obviously broken or malfunctioning elements of their buildings.

The dilemma of modern office-building managers is that they stand perched at the fulcrum of a metaphorical seesaw. One end of the seesaw is heavy with the complaints of occupants and with reports of building performance problems. The other end is high in the air, lifted, as it were, by the silence of building workers who are resigned to building conditions and by those elements of the building that do not need immediate attention. These managers heedlessly pitch their resources at the heavy, dipping end of the seesaw: reduce the complaints, fix up the building, straighten out the imbalance. Their hope is that eventually the low end will lift up, the complaints will slow down to a few, well-placed hints when things in the building go awry, and the work environment will improve to a point where neglected repairs and maintenance work on the building can be undertaken.

This is like running very hard to stay in the same place. Managers know that *reacting* to complaints constitutes more of the disjointed and incremental design decision making that created most of their problems in the first place. Their dilemma lies in how to use their principal gauge of poor building performance—users' complaints—as a *tool* for actual environmental improvement, based on a rational and integrated planning process that is neither reactive nor disjointed. Managers should be able to use occupants' feedback systematically to help them create the optimum work environment.

This dilemma poses many questions, not only to building managers but also to office workers and their supervisors, and to building owners and developers concerned with the initial decisions about an office building's size, cost, and design. Among the important questions raised are: do building managers understand both their building and the people using it well enough to create an optimum situation for both? What kinds of information do office occupants need to communicate to building managers in order to get relief from environmental problems? And how should building operators apply such information to decisions about improving building operation?

STANDARDS OF ENVIRONMENTAL QUALITY

People making environmental quality decisions as managers of a building, or members of a design team, or supervisors of a work group, are not only prejudging what quality is but are also assuming how much it is or will be worth to building users. Although each user has a sense of how much effort to exert to achieve a personally acceptable level of environmental quality (some people will telephone a neighbor to reduce the noise from a late-night stereo; others will live for years on a busy street corner in an urban center lulled to sleep by the sound of the traffic), decision makers do not know how much environmental quality to provide for the people for whom they are responsible.

Bearing in mind that degree of quality is largely but not exclusively a matter of affordability and cost, what kind of guidelines do we have, first, to formulate the right kinds of trade-offs between dollar costs and human benefits, and second, to make the right decisions on how to spend the limited money at our disposal? There are some established ways for making decisions such as whether to improve lighting or analyze air quality, whether to landscape the outdoors or provide sound covers for the computer printers. These are analytic problem-solving models of decision-making, which separate the problem into its component parts: namely, the building (its physical features), and the users (their behavior, needs, and satisfaction). For the purposes of this discussion, these ways of solving building problems are organized into four categories of problem analysis:

- the application of standards and codes;
- the satisfaction of users' needs;
- the reductionist approach; and
- the environmental determinism (or stimulus-response) model

Applying these approaches solves certain problems and generates interesting information about the user-building relationship:

it does not always provide managers with quick and useful ways of assessing environmental quality in their own buildings. A discussion of the uses of each approach, and also what each has failed to do, follows.

APPLICATIONS OF STANDARDS AND CODES

Building designers and managers rely heavily on standards of comfort, health, and safety to ensure environmental quality. Standards indicate how much fresh air should be introduced into an office building's ventilation system; they prevent polluted air from cafeteria exhaust or underground garages from being recycled through buildings; they indicate how much light people need to see to do their work; they help protect workers from excessive noise; they monitor important structural items that prevent walls from cracking and roofs from leaking. Other standards address energy conservation and can be applied to make a building perform energy efficiently. The implementation of health and safety standards and of building code requirements is often seen as *insurance* to the decision maker (building designer, owner, manager) and as *assurance* to the user (worker, occupant, consumer) that environmental quality is monitored in that building. However, in spite of the considerable thought that has gone into the establishment and setting of formal standards, most office-building users know from personal experience that working in supposedly acceptable environments is a far from high-quality experience.

The most prevalent kind of standards are building standards and the directives enshrined in various building codes. Some pertain to the health and safety of building occupants, for example by addressing fire exits and fire control features, or the location of toilets and kitchens, or the width, height, and depth of stairs. The actual form taken by a building is often as much a product of codes and standards (and the inspectors' decisions on how well they have been applied) as it is of design ideas, site conditions, economics, and local politics. Designers and builders find codes and standards frustrating. Code requirements, often formulated without reference to variation in contextual conditions, have to be applied in absolute terms regardless of cost and regardless of the effect of adding to the overall cost of the building.

Suppose, for example, that a small charitable organization is converting a large old mansion into a group home for emotionally disturbed adolescents. They may have to add bathrooms, fire escapes, and a sprinkler system to meet code requirements. These are basic and reasonable additions. But a large discretionary area exists concerning how much more fire protection, for example, such a house may need. The inspector may insist on enclosing the grand old wide central stairway of the mansion and on building an enclosed fireproof stairway. This may cost more than the charitable group can afford. It would make it more difficult for staff to supervise residents on both floors, and it would destroy the aesthetic of the old building, whose elegant homey environment the charitable workers had hoped to use in their work with the needy living group. The fire inspector is not answerable for quantifying the degree of additional protection the users will receive for this additional costly renovation. No one knows whether the reduced risk from fire is worth the extra money, the lost ambience, or the entire project, if the owners have to give it up. In the design of new buildings, as well as renovations, similarly unexpected discretionary judgments by inspectors on the application of code requirements, while unlikely to knock a project the size of a new office building entirely off the rails, can easily result in large additional costs. When unexpected items are added by the demands of building inspectors, other "extra" design features, such as nice landscaping or innovative lighting, which might ultimately protect the quality of the environment for the users, are the first to be dropped.

The invocation of such sacred concepts as "health" and "safety" causes a lack of constraint on the proliferation of standards. Moreover, there is a wide range of discretionary judgment regarding the manner in which health and safety standards are actually applied to building design. In addition to legal and quasi-legal building codes, there is a growing number of organizations devoted to formulating standards—ASHRAE, ASTM, NIOSH, and CSA, to name a few. These standards, if not specifically aimed at buildings, do have an impact on building design and construction through the regulation of construction materials, tools, and procedures.

Certain building standards have been developed to ensure human comfort in buildings. Their existence implies recognition that there is more to protect in a good-quality environment than health and safety. Comfort standards for building interiors, such as those pertaining to lighting levels, heating, and ventilation, claim to ensure that a high-quality environment is a comfortable environment. The ASHRAE organization, which develops such standards of comfort and acceptability for building interiors, represents building professionals who design and operate buildings and who have a need for standards of building acceptability. Using a traditional engineering approach, the comfort research that is sponsored and/or accepted by ASHRAE has tended to favor the use of technical measuring instruments, quantifiable data, and cut-and-dried answers to the puzzling questions of why people react to different physical environments in certain kinds of ways. As testing instruments become more refined, and buildings are built with more advanced technology, ASHRAE standards need to become more precise. In order to do so, ASHRAE research papers usually incorporate some reference to people's behavior, their attitudes, their comfort, and other psychological concepts. But these papers are written by engineers, not psychologists, and their results are often based on experiments that would not be acceptable as behavioral research. Nevertheless, these results, in the form of readily applicable, quantifiable standards of building performance, are accepted as the "last word" by building professionals, who, with little prompting, apply them to building design and equate this with a guarantee of environmental quality.

In applying these standards, building designers and managers are assuming that increased worker comfort will lead to increased productivity and better worker performance. However, this may not be true: increasing people's comfort beyond a certain point may send them to sleep. There is much psychological evidence that people perform better in a *less* than optimum environment, as part of their own struggle, as it were, for improvement. A very much less than optimum environment will, of course, reduce workers' performance. The manager's dilemma is not solved by imposing standards on an environment in order to ensure a higher-quality human experience. Standards of comfort, health, and safety do not identify the delicate balance between optimum worker performance and level of environmental acceptability. Designers and managers who try to increase users' comfort through meeting standards are likely to be frustrated.

SATISFACTION OF USERS' NEEDS

Studies of the people-building relationship have tried to analyze and understand how to design optimum environments for people through evaluating existing buildings. Much of environmental evaluation relies on people's rating of their own satisfaction as the criterion of a successful environment.

Building evaluation studies, the findings from which will be discussed in chapter 3, analyze ways in which the physical environment does or does not meet people's needs. Criteria for better building design have emerged from building evaluation research; they differ from comfort standards and building codes in that they address levels of human need other than comfort, safety, and health, such as social behavior and psychological satisfaction.

The user-satisfaction approach reflects the beliefs that the best buildings are those with the most satisfied users, and that to be good, environmental design should meet specific human needs (Vischer 1985). This assumes both that needs in buildings are constant and can be identified and that design is capable of meeting them. By creating a direct relationship between the number of needs and the number of need-meeting or user-derived design features in a building, a sort of hierarchy of user requirements is implied, as illustrated in figure 2-1.

The diagram shows that a building may meet the broad-based human requirements for safety (at the base of the pyramid) and be habitable. If it also meets standards of comfort and satisfaction (at the apex of the pyramid), because they are the most personalized and therefore the most difficult to satisfy, it is more than habitable—it is a high-quality building.

2-1. The habitability pyramid.

This is a powerful model, used implicitly by many designers, researchers, and managers in making decisions about building quality. A building manager may ask a group of office workers what they need the most, and he will be told that they need "more space," "more privacy," and "more storage." In order for the building to meet these needs, major reorganization of space is necessary, and some other desirable aspects of the environment may be lost: for example, in giving workers more individual work space or storage, their lounge, lunchroom, or conference room may be lost. In giving them more privacy, their social contact, their views out of the windows, and some circulation space may be lost. Many office users may be unprepared for these trade-offs. Reorganizing the space may generate a new set of unmet needs that mean more headaches to the building manager.

It may therefore not be useful to define human needs as finite, quantifiable units to which something as concrete as a building can respond. The possibility that an environment is responsible for meeting people's needs while they sit around waiting for their needs to be met is an unfortunate side effect of the user satisfaction approach. The most carefully designed environments often do not meet users' needs in the ways anticipated, so designers and managers may decide that they have failed, or that building users are a "bottomless pit" of needs that no building will ever meet.

It is more useful to decision makers to acknowledge the limitations of this model and to recognize that the shortcomings of this approach are similar to those of the ASHRAE and standards model: neither model incorporates the interaction of users with their environment. People are not passive bundles of needs waiting to be met or recipients of a comfort-providing building design. People seek comfort from their environment; they change their environment, adapt it when they can, and adapt to it when they cannot.

THE REDUCTIONIST APPROACH

The reductionist approach attempts to isolate units of environmental experience that, decomposed, can be observed and possibly measured, and composed, constitute environmental quality. This approach applies the conventional scientific paradigm of analysis (decomposition) of the user's experience of the building. This is usually done through measuring users' *perceptions* of their environment. Students of the approach assert that such perceptions, in order to be reliable, must distinguish between personal preferences, which are subjective and not reliable, and comparative appraisals, which are more reliable because the individual is judging the item in comparison to an implicit or explicit standard of comparison (Craik and Zube 1976).

Being able to understand and make use of users' perceptions of their environment could be useful to both managers and designers of buildings. However, it is difficult to do this systematically, in a simple and manageable way that does not bog down in uncontrolled subjectivity. Researchers who have used this approach to measuring environmental quality, usually by using a simulation of the environment to be assessed, have developed scales for respondents to use to measure their perceptions of predetermined environmental qualities. An example of this is the PEQI, or Perceived Environmental Quality Indicator, a set of standardized indices that can be applied to physical features perceived by building users to contribute to their experience of environmental quality (Craik and Zube, 1976). PEQIs, like most other tests of perceived environmental quality, are derived from semantic differential tests of various (outdoor) environments. The semantic differential test pairs words with opposite meanings as descriptions of an environment, and users rate the environment as closer to one end or the other, as in "beautiful-ugly, uncrowded-crowded, quiet-noisy," etc. The ultimate but unrealized aim of the reductionist approach is to correlate people's ratings or indices with objectively measured attributes of the environment they are evaluating.

For decision makers, the fact that the measure of human perception is abstracted from the environment being measured makes the reductionist approach impractical for solving building problems. The indices can be used to *infer* quality, but the quality itself remains mysterious, amorphous, abstract. Because the environmental quality indices do not define quality, they cannot be used as a basis for making decisions to effect environmental change any more effectively than the other analytic approaches. Using the reductionist approach, the designers and managers cannot plan for environmental improvement on the basis of scale ratings that have not been validated by measures of building performance. They therefore fall back on the more practical building standards approach.

ENVIRONMENTAL DETERMINISM

Environmental determinism assumes that people's behavior in buildings is directly caused by the physical elements of those buildings. Therefore, if people's behavior is unsatisfactory—they are complaining, they do not get along, their work is not getting done—the solution to the problem lies in changing the design of the building. In this approach, the building and the environmental elements that it comprises are considered the *stimulus,* and what people do in buildings is seen as the *response.*

The BOSTI study mentioned in chapter 1 identifies environmental satisfaction, job satisfaction, and job (or worker) performance as behavioral effects that are "caused" by the physical environment (Brill et al. 1985). The study argues that elements of office design determine increases or decreases in these three effects. It identifies specific environmental elements such as lighting, noise protection, and furniture design, which can be manipulated to improve people's satisfaction and job performance. One of the often-cited examples from this study is the improved job performance—especially among high-salaried workers such as managers and senior executives—resulting from greater enclosure of the work space. By applying "economic" formulae to measures of worker productivity, the authors compute the dollar gains in productivity from improvements to these environmental features. So if a highly paid technical specialist is given a fully enclosed office, not only does her work improve, but the better work she performs is more valuable because of her high salary. In this particular form of environmental determinism, the changed environment results in a changed behavior that contributes to the assets of the organization.

In this model, environmental quality is defined as dollar increases in output based on a person's salary as compared to dollar investments in interior design. Building designers and managers who apply the deterministic model impose a heavy burden on themselves. They have to prove, somehow, and rarely satisfactorily, that there is a direct and tangible payoff from improving the quality of buildings: for these building managers, the desirability of environmental improvement is not sufficient for its own sake. By trying to *demonstrate* that physical changes to a building cause measurable increases in user performance and satisfaction, managers and designers are knocking their heads against a wall (Sommer 1983). They need only to point out something that everyone knows: that a building that works for its users is a priori a better environment than one that does not.

As long as decision makers view people using buildings as people responding passively to stimuli, they will not be able to apply environmental determinism to solving building users' problems with any greater success than the other approaches. Environmental improvement does result in money saving, productivity increasing, and health preserving, but only when the physical environment and the users' interaction with it are treated as a system, the building-in-use system, which is being improved.

THE NEED FOR A MORE DYNAMIC APPROACH

One of the conceptual difficulties with the "habitability pyramid" is defining who a building's users are and what they require from the building. Different users groups want different things from a building. Users are not limited to building occupants; building users also include the clients of the architects and engineers, those who apply the economic formulae that determine building form and who pay the designers' bills. These users require at least economy in construction. For them the building is a commodity. The client during the design and construction process may not be the ultimate owner of the building, and the owner is not always the occupant. Owners and occupants are building users at a later stage. Owners may be more concerned, for example, about the building's energy costs than either the building's tenants or the architect's client.

Other building users have ongoing maintenance and management tasks that keep the building in operation. Their experience of the building differs from that of the people who come in to work there every day. They also have different criteria for a good building—for example, that it should be easy to keep clean and in good repair. A fourth group of building users includes those who live and work nearby, those for whom the visual experience of the city is shaped by the presence of that building. Their requirements have little to do with energy costs or cleaning, but they have a lot to do with appearance, access, and form.

In trying to define who the users are, and pointing out the different requirements they have of a single building, it is clear that it is as impossible for one building to meet with equal effectiveness some of the needs of all users as it is for it to meet (equally or not) all the needs of a single user group. It is also evident that the requirements of different user groups are not always likely to be compatible. Indeed, the requirements of one user group (e.g., the need for easy cleaning and maintenance) may actually conflict with the requirements of another group (e.g., the need for soft, light-colored, sound-absorbing finishes).

Given these stringent and conflicting requirements, it is not surprising that most buildings generate more complaints than praise from users. Moreover, the diversity of user expectations makes the notion of building improvement complex and difficult: what is improvement to one group may not indicate improvement to another. In such a contradictory context, it is not surprising that the static analytic approaches to the question of what to do with and about users' complaints are not adequate to solve building problems. It is not clear how the analytic approaches will provide *either* a useful and practical definition of building quality, *or* a method for achieving environmental quality in less than adequate building environments.

In summary, while there is a psychological research tradition of human need definition that has been applied to the study of buildings and from which quite a literature has developed, there is clearly a difficulty in defining the people side of the people-building equation as long as people are seen as a set of needs rather than as adaptational forces in the building-user system. The invocation of building standards for safety, health, and comfort asserts that if specific physical features or conditions are present, then ipso facto human safety, health, and comfort are assured. The fact remains that many workers feel uncomfortable and dissatisfied in office buildings where most standards are met, but it is not clear whether this discrepancy has to do with poor standards, with the lack of consensus among user groups, with an overemphasis on worker productivity and the limitations of cost-benefit analysis, or with the failure of modern office buildings. In short, the analytic approaches alone do not provide a reliable indicator of the meaning of *environmental quality.*

ENVIRONMENTAL QUALITY

Quality in office buildings is that attribute of actions and things that, if present in office buildings, reassures not just those who work inside them but also the owners, developers, and managers of such buildings and the people in the street who watch them being built and sit in their plazas and atria. Quality is something more than increased productivity of the individual or increased profit of the organization. To define quality is not an easy task, but speculation on the meaning of quality is necessary before advancing a definition of "environmental quality." In *Zen and the Art of Motorcycle Maintenance,* Robert Pirsig wrote that

> Quality is the continuing stimulus which our environment puts upon us to create the world in which we live. . . . Now, to take that which has caused us to create the world, and to include it within the world we have created, is clearly impossible. That is why Quality cannot be defined. If we define it, we are defining something less than Quality itself (Pirsig 1975, 245).

We are quick, however, to recognize the *lack* of quality in something. Poor quality or lack of quality inspires criticism and rejection. The difficulty is *not* in knowing quality or in recognizing it where it exists; the difficulty is in assessing quality, in measuring it and objectifying it so that it can be described to others. Although the experience of quality is subjective, it is also consensual. This does not mean it is capricious or transient; this means it is real.

> At the moment of pure Quality perception, or not even perception, at the moment of pure Quality, there is no subject and there is no object. There is only a sense of Quality that produces a later awareness of subjects and objects. At the moment of pure Quality, subject and object are identical (Pirsig 1975, 284).

From the point of view of those experiencing the quality, the fact that there is a "later awareness" of quality means that the presence of quality *can* be defined—and communicated—to those who have not had direct experience of it. In fact, the experience of environmental quality—of working in a good-quality environment—can be communicated to those who make decisions about buildings. Designers, building managers, and office workers' unions, for example, may make an assessment of the quality of a particular environment to aid them in the next design, to help them set priorities on building improvements, or to intervene on behalf of their members.

Environmental quality is the combination of environmental elements that interact with users of the environment to enable that environment to be the best possible one for the activities that go on in it. Environmental quality in one office will not necessarily resemble environmental quality in another in every detail, but there are constants across office environments that can make the difference between a good- and a poor-quality environment in almost all offices. Environmental quality is a question of degree: most offices can be a degree or more better than they are now. The question is not *why* improve office quality: people simply work better in a better-quality environment. As most of the degrees of improvement (discussed later in this book) are small-scale and inexpensive, the question is really, *why not* improve office environmental quality?

THE PSYCHOLOGICAL DIMENSION OF THE BUILT ENVIRONMENT

Few people, when asked, will deny that they have a right to quality—quality in all things, including environmental quality. As individuals, we like to feel we have access to the best-quality lives we can afford, and if we cannot afford the quality we think we deserve, we try to "get ahead" and to work more successfully in order to afford more quality. Collectively, however, we are more often in the position of having others make decisions about the quality of our environments. For example, if an apartment building that we do not like is approved for construction on the street where we live, we realize that someone at City Hall is making a decision that directly affects the quality of our living environment, and we may even be moved to register a protest. Similarly, in the buildings where most people work, supervisors, building engineers, and property managers make decisions in the interests of the collective that determine the quality of each individual's work environment. A typical issue is the temperature setting of the thermostat that regulates a group work space. Set to meet the needs of the group, it is perhaps too warm or too cold for an individual user. In making such decisions, decision-making individuals responsible for building quality are setting, inadvertently or not, standards for the work group and the organization.

One of the reasons why setting environmental standards for a group of users is difficult is that "objectively quantifiable" building standards do not take the psychological dimension of building performance into consideration. Windows in buildings provide a good example. In the large deep-floor office space that is typical of modern office buildings, workers receive more and usually better ventilation from a conventional mechanical air distribution system than from opening windows at the perimeter. In the name of safety, easier maintenance, and energy conservation, many modern buildings have sealed windows. People who work in sealed buildings tend to dislike them, and they complain about poor ventilation, in part because they cannot open the windows. Enabling windows to be opened may make an indistinguishable contribution to air quality in the building, but because of the psychological importance of windows, this design decision may significantly increase the *comfort* experienced by the people working there.

The ASHRAE approach argues that opening windows does not measurably improve ventilation and reduces HVAC performance efficiency by reducing building pressure. The building-in-use approach argues that the reduction in efficiency will be more than offset by the increase in people's psychological comfort and enjoyment of the building. The psychological dimension of building use is integral to the definition of environmental quality. The building manager makes good decisions in the interests of the collective of office users by incorporating the psychological dimension of building use.

IMPROVING THE OFFICE ENVIRONMENT

The psychological dimension of building use is central to the building-in-use approach. To define quality in building environments, the physical environment and the users who occupy it must be considered as a whole, or a system, and not conceptually separated into subject (a user with needs) and object (a building with standards). A recent symposium on the impact of the work environment on productivity emphasized the need for a new way of defining a building, not as an entity separate from the people in it, but as part of a building-and-users system that incorporates the effects of the environment on the users and the effect of users behavior on the environment:

> The building process is increasingly perceived, not as a one-shot supporting process of producing "building as product," but as a continuous, complex and parallel supporting process of managing change in an attitude of "building-in-use," responsive to the changing needs of the core activity, the work process (Dolden and Ward 1986, 387).

This new paradigm of the building-in-use involves a redefinition of the knowledge base and information needed to understand buildings, and a redefinition of the actors and organizational systems needed to produce and improve buildings. Thus, improving the quality of the office environment is improving this system, the system that is the occupied environment, the building-in-use. Figure 2-2 compares the two approaches to effecting environmental improvement discussed in this chapter. Environmental improvement means increasing the quality of the office environment: the notion of environmental quality is an integral part of the concept of environmental improvement. Improving the environment means improving the quality of working life.

"Improvement" is a process of directed change, based on appropriate use of the right kind of knowledge, and with an awareness of context that allows for the control of unwanted side effects (de Zeeuw 1981). Actions for change can either decrease or increase quality. In order to know whether an increase or a decrease has occurred as a result of an "improvement," it is necessary to define a variety of underlying criteria against which to evaluate change and improvement. Such varying criteria must incorporate differences in the perceptions and judgments of the building's various user groups: in short, the blending of subject and object that is the experience of quality (de Zeeuw 1980).

The remainder of this book illustrates one way in which

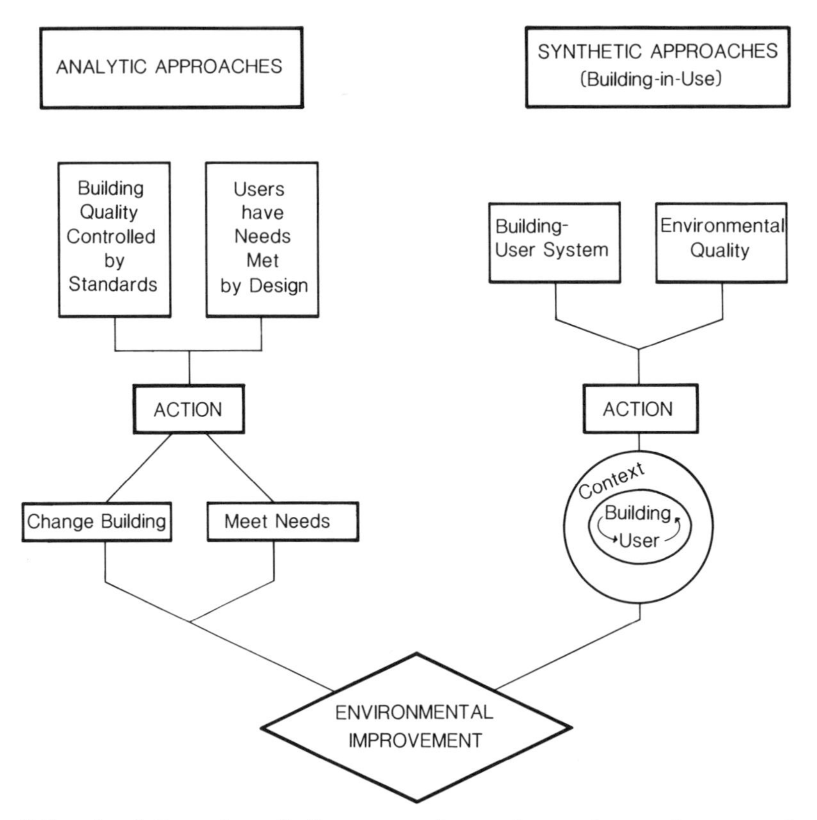

2-2. Analytic and synthetic approaches to improving environmental quality.

environmental quality can be assessed and environmental improvement decided upon, using the blending of subject and object, user and building; using underlying criteria of user consensus; and using an awareness of the context of the problem. The book shows how the systems model, of which the central concept is the building-in-use, is more effective in solving problems of poor building performance than the analytic approaches. Our argument is that the concept of environmental quality is an objectifiable one; that people's perceptions are a measurable reality of environmental quality, and that the building performance measurements described in chapter 4, while helpful in explaining technical aspects of building performance, do not provide a complete picture of building quality for the person working in the same building every day.

Designers and managers need a way to integrate systematically the information on problem definition yielded by users' complaints, the technical building performance information that provides solutions, and their own operating constraints such as budgetary considerations and the management philosophy of the organization. Most managers know that environmental quality does not exist outside the context of users' perceptions. They know that improvements in environmental design will substantially increase worker productivity, if the improvements selected correspond to users' perceptions, to the extent that these can be objectively measured. The way to know which improvements to effect is to understand and tap into occupants' experience of environmental quality. This involves measuring not just technical aspects of building performance but also the environmental perceptions and sensitivities that color workers' perceptions of quality, and, most importantly, doing so by measuring and assessing these as an interacting system: the building-in-use.

SUMMARY

Owners and managers of buildings sooner or later need information about the quality of the building stock for which they are responsible. On many occasions this need develops in response to complaints from building occupants or in reaction to some unexpected incident of building failure. For many people involved with and responsible for buildings, there is ongoing curiosity about and interest in knowing how well they are performing; but only gross indicators like an absence of occupant complaints and low operating costs provide them with a sense of their own success at building management.

Certain dilemmas face building decision makers—managers, space planners, owners—when they attempt to improve the environmental quality of their buildings. Many of their efforts at building improvement do not "pay off" in terms of increased satisfaction or productivity of workers. Although managers' main source of information about poor building performance is users' complaints, conventional ways of defining the problem encourage the separation of the building and its performance from the ways in which it is used and experienced by the people in it.

A review of four analytic approaches to understanding buildings is available to decision makers, although much of it is a user-building problem adequately enough to provide solutions. People responsible for solving building problems need to integrate the psychological dimension of building performance with information from other sources and to understand the interaction of building and user and their mutual adaptation.

Before moving on to details of the building-in-use method, which constitute chapters 5 through 7, we need to understand what is already known about office environments and how they affect people. A rich literature on the evaluation of existing buildings is available to decision makers, although much of it is based on one or more of the analytic approaches described above. Building evaluation studies are the topic of the next two chapters. Chapter 3 reports the knowledge we have from evaluating office buildings from the point of view of the behavior of building users. Chapter 4 describes the building performance approach to office building evaluation. Some of the insights required to improve not only building quality, but the quality of the work that goes on inside buildings, will no doubt occur as more everyday knowledge is accumulated and shared on the ways that buildings affect our lives.

Chapter 3

OFFICE BUILDING EVALUATION: SOURCE OF EXISTING KNOWLEDGE

What is building evaluation? The aim of systematic evaluation of a building (including an office building) is to acquire knowledge. This increases our ability both to solve immediate building problems, and to effect improvement over the long term. The knowledge accumulated from evaluating buildings from the point of view of how they are used by people provides us with an understanding of effects of the built environment on people's behavior. When an office building is studied through the experience of its users, knowledge is generated both about the performance of the building and ways to improve it, and more generically about how people's work and social behavior is affected by features of the building environment.

In this chapter, we review some studies of office building utilization in order to learn more about prevailing methods and models of person-environment interaction. Some of these studies apply the analytic approaches described in the previous chapter. Others take a more dynamic "systems" view of people and buildings as interactive and mutually adaptive. The relative merits of these two approaches and the quality of knowledge they have yielded is the topic of this chapter.

WHY EVALUATE BUILDINGS?

Organizations evaluate their buildings for two purposes. One purpose is to gather information that will enable a building problem to be solved. If some workers in a building are complaining about headaches, for example, some information on whether these are caused by bad lighting, inadequate ventilation, or something that is not connected to the building will help the building manager determine how to solve the problem. Another purpose of building evaluation is to acquire generic knowledge about the relationship between people and buildings. In this case, while information on problems may also be acquired, knowledge about buildings will accrue only if the evaluation is oriented to more than immediate problem solving. Organizations with an interest in acquiring information to solve problems which also want to amass knowledge about building use and generic improvement may implement an "environmental design evaluation program." Their hopes are:

1. To extend our understanding of human behavior by further documenting the transactions of people and the built environment.
2. To extend the design process to include evaluation and the development of a feedback mechanism for the inclusion of evaluative data in the making of design decisions, both for fine tuning existing environments and creating new ones.
3. To provide an important body of data for use in the education of future design professionals and for use in continuing education programs.
4. To obtain the kinds of data required for the analysis of the efficacy of public policies and programs which support and constrain the design of a range of environmental settings.
5. To begin to develop a capability for the prediction of user satisfaction and environmental fit for environmental impact assessment in its broadest definition (Friedmann et al. 1978, 5).

These are all good, sound reasons for going out and evaluating buildings, and all have at one time or another been invoked by designers, researchers, and building managers. However, they express objectives akin to social welfare, motherhood, and apple pie; they do not provide a building owner or manager (or an architect or a labor union) with cut-and-dried cost-benefit reasons why a building should be evaluated. The use of evaluation to solve an imminent building problem is a more obvious tool for managers.

Office building evaluation studies have drawn their momentum from three pervasive problems of white-collar work environments: 1. Those causing physical health hazards; 2. those affecting people's relationships to each other and to their environment; and 3. those that impede the work effectiveness of individuals and groups (Davis and Szigeti 1986). These problems emanate in large part from three main sources:

- the large-scale impact of the "open-office" concept, which was introduced to North America in the '50s and '60s, an impact that remains controversial in spite of the widespread implementation of the concept;
- the recent rapid introduction of electronic technology into office work, and its impact on work organization and the work environment;
- occupational health concerns regarding the effects of modern office-building interiors on worker health and well-being.

The next section looks closely at each of these issues and the research, analysis, and knowledge acquisition that has gone into trying to resolve them.

THE OPEN OFFICE

Originally introduced to North America from Europe by Dr. Rod Planas and the Quickborner office planning team, the concept of "open-plan" office design was marketed across the country as a new cost-effective and flexible way to plan office interiors (Planas 1978). The design principles of open office landscaping, or *Bürolandschaft* were introduced to make the open office a workable, aesthetic, and inexpensive alternative to traditional office layouts.

In fact, what is called *open-plan* today often bears little resemblance to the principles of *Bürolandschaft.* In a typical contemporary open-plan office, one is often treated to a forest of so-called acoustic partitions or screens that enclose each desk. These cut off the natural light entering from perimeter windows, segregate and isolate workers without providing much acoustic or even visual privacy, and impede adequate air flow through the interior space. They create "cubicleland" for workers (Stone and Luchetti 1985). The plants, spacing, and other elements of the original open-plan concept are often not in evidence in a typical office. This explains, in part, why office workers in North America often trash the open-plan concept; the most common complaints are high noise levels and lack of privacy.

The open plan continues to be a favored approach to office planning because it is flexible enough to accommodate the frequent moves that are typical of the modern office, and because it is cheaper and more efficient when designing a new structure to service large open floors than to plan cellular interiors. Proponents of open-plan office design maintain that it facilitates interpersonal communication. An important modern component of open-plan layouts is the proliferation of furniture systems, where modular components, which include desks, shelving, and acoustic screens, are fashioned into a variety of work-group configurations in an open floor and are changed as necessary while retaining a pleasing appearance of uniformity and efficiency. Many organizations spend large sums of money on streamlined and elegantly colored furniture systems in the belief that these represent solutions to the problems of the open plan. For example, the acoustic partitions provide voice privacy, the desks have ergonomic dimensions for the comfort of workers, and the shelves conceal task lighting that illuminates the work surface. However, not all problems of office layout are solved by furniture systems, especially if the open plan is not well designed to start with. They represent a considerable cost outlay to an organization without always yielding apparent large benefits in improved employee performance or satisfaction.

If landscaped open-office space were properly implemented according to the design principles of *Bürolandschaft,* it would be equal to or more costly than cellular or enclosed layouts, and as difficult to reorganize or change. For example, a key factor in *Bürolandschaft* design is adequate spacing of individual work stations. Distance and orientation of desks can control many of the acoustic problems that are commonly reported by open-plan office workers (Harris, Palmer et al. 1981). This spacing provides better privacy but consumes more space on the floor. In such a scheme there would be as few or fewer work stations per floor than in a cellular layout.

Studies have suggested that workers in open plan are less productive and less satisfied than workers in enclosed offices (Hedge 1982; Sundstrom et al. 1980; Oldham and Brass 1979; Boyce 1974; Brooks and Kaplan 1972). As the reasons most often cited are noise and intrusions, reduced job satisfaction, and inadequate privacy, these authors and others recommend more individual enclosure to increase worker productivity. The conclusions of these and other studies have led office managers to discredit the open-office concept, and some organizations in Europe and North America are imposing a policy of cellular interiors in their office buildings.

Other concepts currently replacing *Bürolandschaft* are *Gruppenraum* or group work space, and the *office village*. Enclosed office layouts should not be considered the only alternative as they often have other costs, such as long interior corridors with rows of doors, a monopoly of the perimeter offices over window access and view, and inadequate ventilation. For most modern organizations some form of open-plan office is a practical necessity to handle the speed of organizational changes and the rapid introduction of electronic equipment. This is usually combined with some enclosed spaces for senior executives, managers, and meeting rooms.

OFFICE AUTOMATION

The recent introduction of electronic equipment into modern offices has posed temperature, acoustic, illumination, and ergonomic problems. Some building evaluation has focused on the impact of installing new technology in existing buildings and the impact of extra power, cabling, and heat-load requirements on a building's structure and systems (Duffy et al. 1983). Computers themselves have stringent environmental requirements, such as cool and stable temperatures, adequate humidity, and protection from dust and fibers, which make major demands on office space design and layout. Unlike human beings, the computer does not adapt environmentally, and its demands must be met in order for it to function.

Even microcomputers and desk-top terminals make demands on the environment that are often exacerbated by the random and fortuitous way in which electronic technology is introduced into most offices. In spite of glowing accounts of the computerized "office of the future," the reality is that various electronic components are not introduced all at once as part of a carefully planned changeover (Kleeman 1986). Rather, as electronic equipment is added piece by piece, more cables fill the already inadequate wiring space in the walls and ceiling, printers are placed in corners where they are hard to find but where their noise is most remote from other workers, and people put their new terminals on their desks, often by the window or under a light that causes glare from the screen and visual difficulties over the long term.

According to the increasing number of books, papers, and learned articles on the subject of office automation, planning solutions are available to organizations that are shifting to electronic technology (e.g., Cohen and Cohen 1983). Many focus on the physical and spatial correlates of work reorganization resulting from the introduction of automated equipment (Gregory 1983; Rubin 1984; Becker et al. 1986). Others address the impact of the changing office interior environment on worker behavior; for example, the acoustic impact of printers and keyboards, the problems of glare on screens, and the comfort of furniture, and screen and keyboard placement. Most agree that the impact of office automation is complex and far-reaching and needs to be handled differently in new buildings from how it is handled in changeovers in existing buildings.

The field of *ergonomics*, which had its inception in the

design of work space and equipment for efficient human use, has focused on the design of the microenvironment of the individual worker in the office (Murrell 1969; Coe 1983). The design of man-machine systems in industry has led in recent years to specifications for office furniture and space design that have grown in importance with the introduction of electronic equipment and repetitious, machine-based tasks in offices.

Much current research in "human engineering" or "human factors" is oriented to the specification of standards or guidelines against which existing furniture systems and spatial layouts can be assessed (e.g., Cakir et al. 1980). Improperly designed furniture and lighting for video-display terminals (VDTs) are now considered major threats to worker health both because of the extended time periods office workers spend at screens, and because of the increasing number of workers using screen-based equipment. More unions are negotiating rest periods for workers at electronic terminals. Thousands of people who work every day at video-display terminals in modern offices are exposed to health risks, some of them serious, ranging from improper lighting, to muscle injuries (e.g., Repetitive Strain Injury), to microwave radiation. As three out of four office workers will be using computers by the year 2000, and of those who have them now, more than 20 percent use them more than five hours a day, the incidence of VDT-related problems, known in the media as "VDT Syndrome," and in Japan as "Technostress," is likely to increase. Thus, the need to acquire a better understanding of the impact of the electronic environment on the office worker is an urgent one.

THE OFFICE AS A HEALTH RISK

Although an increasing number of office workers are using electronic equipment, there are still plenty who do not. A large number of men and women whose "crime" is nothing more than going to work in an office every day are running serious health risks from building-related sources other than electronic equipment. These include indoor air pollution, glare from lights and windows, and stress from noise (Stellman 1977). Decisions taken in the late '70s and early '80s to make office buildings function more energy efficiently have in many cases generated direct threats to worker health and comfort. For example, restrictions on air circulation and the tight sealing of building openings and air leaks have caused occupational health specialists to identify *sick-* or *tight-building syndrome.* This syndrome is characterized by a set of symptoms reported by office workers, including fatigue, nausea, dry skin, and sore throats (Carlton-Foss 1983; Finnegan et al. 1984; *Newsweek* 1985). The Terrasses de la Chaudiere is a case of sick-building syndrome.

Fluorescent lights fade with the passage of time, and failing lamps often give rise to flickering and buzz. These factors cause headaches and eyestrain, as do poor lighting conditions like glare, veiling reflections causing poor visual contrast conditions, and overlighting. Tiredness and headaches in the afternoons—problems reported by many office workers—have been linked to high levels of carbon dioxide arising from insufficient air changes per hour (Makower 1981). New floor coverings, drapes, and other office furnishings can emit potentially toxic gases, as can numerous other seemingly innocuous office features such as photocopying machines and janitorial supplies. Energy conservation in office buildings means that less air is circulated by the mechanical ventilating systems, and potentially toxic chemicals in the air are not rapidly flushed out of the building (Vischer 1987a). Low humidity causes dry skin and lips. Windowless offices can cause depression and disorientation.

Even if human health is not directly affected over the short term by such building conditions, stress-related symptoms have been traced to bad acoustic, illumination, and thermal conditions in offices (Craig 1981; Evans 1982). Computers are often better treated than people, being housed in carefully monitored spaces where temperature and humidity are controlled to within a very narrow range, and not just anyone can barge in and invade privacy. Workers' experience of environmental stress is often a source of low morale and decreased productivity.

Given what is known in these three problem areas of office design and planning, what can building evaluation studies

portant (as some have inferred), b
cal dimension of man's relationsh
tional physical measurement. Imp
the importance of psychological
analysis of any behavioral phenom
and productivity or building-in-us
mental quality. There are "intrinsi
work in people's reactions to and j
just as there are in people's job sa

Later writers who have studied
tions to their physical environmen
Steele (1983) identified the follov
interactive relationship between p
which they work:

1. Shelter and Security
2. Social Contact
3. Symbolic Identification
4. Task Instrumentality
5. Pleasure
6. Growth

He is at pains to point out that the
the building performs for the oc
opportunities for *interaction* betw
environment that can be used posi
or negatively (to block action) b
description of the "Growth" dime
states:

> By Growth I mean any of a number of c
> new skills; achieving a greater sense
> aware of one's personal preferences,
> achieving a greater sense of compete
> understanding more about how the w
> Settings also have an impact on group
> group norms and group problem-solvi

Whereas the objective of Herzb
nature of the relationship between

contribute to our knowledge of how to solve them? Building decision makers must set priorities on whether the case for environmental quality is based on solving the problems of the open plan, on the successful introduction of office automation, or on protecting workers from stress and ill-health. In fact, the case for environmental quality is based on evidence that an improved environment, whether it had "problems" or not, results in better job performance by office workers.

In the next section, we review the history of office environment studies, drawing the distinction between the static, analytic model and the dynamic interactive model of people in buildings that provides the basis for the building-in-use application. The review will indicate ways in which existing knowledge can be applied to solving the generic office environment problems of open-office design, office automation, and occupational health hazards.

SOME HISTORY OF ENVIRONMENTAL EVALUATION IN OFFICES

The dynamic, interactive approach to understanding the psychology of the work environment is not new or recent; it has roots in the intellectual and research heritage of the field of study of human beings and work. One of the first researchers to postulate dynamic interaction between workers and their environment was Frederick Herzberg.

Writing in the 1960s, Herzberg was one of the earliest psychologists to include the "environment" in his taxonomy of work-related factors affecting job satisfaction and work performance (Herzberg 1966). However, the work environment to which Herzberg refers is its *context:* the set of external factors that bears upon the individual in the performance of his or her tasks. These include salary, status, job security, company policy and administration, and working conditions. They are contextual or "extrinsic," according to Herzberg, because alone they do not generate job satisfaction in the worker; they are either

neutral or sources of dissatisfactic
from the "task-centered motivator,"
growth that is experienced "intrinsi
berg does not specify architecture,
visual, and acoustic comfort as con
tions"; his writing implies that the b
work is an extrinsic "hygiene" fa
"company policy and administratio
either neutral or a source of dissa
alone cannot generate job satisfacti

Herzberg summarizes the findir
studies as follows:

> First, the factors involved in producir
> and *distinct* from the factors that led to
> factors needed to be considered, depe
> or job dissatisfaction was involved,
> feelings were not the obverse of one a
> satisfaction would not be job dissatisf
> tion; similarly, the opposite of job dis
> tion, not satisfaction with one's job. Th
> up of two unipolar traits is not unique
> to grasp (Herzberg 1966, 76).

There is no good reason in moder
conclusions to workers' *environme*
an extrinsic factor, workers are eithe
dissatisfied with it. The question
whether or not there is an intrinsi
satisfaction produced by another set
produce environmental dissatisfacti
site is not environmental dissatisfac
environmental satisfaction. If this ar
some of the separate and distinct
satisfaction or dissatisfaction with
covered, then we would have a case

Herzberg made good use of th
Studies (1927–1932), which found
did not systematically relate to beh
He concluded, however, not that

Steele's objective was to make the following important point: an organization should not underestimate the impact of the physical setting on the behavior of its workers. It can, in fact, contribute to its own organizational health by analyzing, understanding, and responding to the effects of its building environment on its own task processes and organizational decision making.

Following from Herzberg's and Steele's thinking is Franklin Becker's analysis of how "workspace" has evolved to reflect changing fashions in management philosophy (Becker 1981). Building on Steele's interactive model, Becker points out that the attention a person or work group pays to (selective) environmental features determines, in turn, the impact of that environment on that person's or group's behavior. He states, "Physical settings are created through a series of attention cycles characterized by . . . information transformation." (ibid., 41) In this way, Becker endorses Herzberg's categorization of the physical environment as an extrinsic factor, which is either not in the worker's consciousness at all (i.e., no attention is paid to it) or is a source of dissatisfaction. He is also reformulating Steele's interactive model: it is possible that selective attention may be paid to a particularly *satisfying* feature of the environment.

To Becker, Taylor's (1911) theory of "Scientific Management," which dominated the organization of work in the first half of the twentieth century, has profoundly affected the ways in which organizations make decisions about the work environment. The standardization of work into discrete units or tasks assigned to each worker on the basis of a standardized level of competence or skill evolved from the study of manufacturing processes and industrial environments. This approach, in turn, gave rise to unquestioned contemporary acceptance of a standardized work environment in which all workers receive standardized thermal, lighting, and acoustic conditions in a minimally sized work space with standardized furniture. Becker's point is that this industrial norm is not necessarily appropriate to white-collar work and work environments; he recommends a "cybernetic" planning model in which information on human needs and changing task requirements is fed systematically back into the environmental planning process. This is how

Becker sees the interactive user-building approach to the acquisition of knowledge applied to the planning and design of new environments and to the solution of building problems.

These writers are among those who are clear-sighted enough to recognize and indeed hail the interactive nature of the person-environment relationship. It is one thing to recognize and write about it; it is another to put it into practice. In the history of this field of study, there has been no acceptable methodological approach to building evaluation based on the systems view of user and building. The lack of a practical building-in-use methodology is attributable to four factors.

The first is a general reluctance on the part of researchers to deviate from conventional social science methodology and its data-gathering and analysis dictates: the analytic approach. The second is that methods that have attempted to capture the "wholeness" of the building-user system are detailed and cumbersome, and do not lend themselves easily to field applications by untrained people (Barker, 1970; Margolis, 1981; Ventre, 1988). The third factor is the predilection of social scientists to analyze and report building users' behavior (e.g., their satisfaction) without reference to the specific environmental elements of the building being evaluated, that is, emphasizing psychological factors at the expense of information about building performance research. And finally, as long as trained researchers perform building evaluation in lieu of the people who actually design, build, own, and operate buildings, they will exercise ultimate control over the information yielded by the studies, trading off its practical applicability in favor of its research integrity. The effects of these four factors on modern building studies are reviewed below.

CONTEMPORARY BUILDING EVALUATION

1. Measuring User Perception: Many contemporary researchers appear to think nothing of reporting the results of surveys of office building occupants with little reference to or description of the physical features occupants are evaluating (Harris et al. 1981; BOSTI 1982; Spreckelmeyer 1985). Data from building users are reported as though they occurred in a vacuum, separate from both the organization and the building they purport to describe. This implies the existence of some generic truth in such findings as:

> People tend to arrange their personal workspace in similar ways (Goodrich 1982, 373); or
>
> Path-finding for strangers, new employees, and visitors is a problem reported by more than two-thirds of all office workers in all kinds of offices—open and closed, large and small, new and old (Brill et al. 1984, 236); or
>
> It is evident that the issues of conversational and visual privacy and space for storing work items are rated negatively in comparison with other attributes of the environment (Spreckelmeyer 1985, 7).

Although not without interest, these declarations cannot be construed as evaluative of a particular building environment if no links are reported between environmental features and those particular attitudes or behaviors. These findings are the result of research into the psychology of users rather than into their interaction with the building.

Studying the psychology of users and their perceptions separately from the environment they are perceiving poses the philosophical problem of whether the existence of a physical element like a table—or a building—can be separate from people's perception of it, a question found worthy of debate by existential thinkers like Husserl, Kant, and Hegel. In practice, it is of dubious usefulness to evaluate a building by separating users' perceptions of it from the building itself. For example, architectural design award programs apply a variety of criteria to judging building quality that often do not include user perceptions, but emphasize appearance, form, and materials. Evaluations of such buildings in use have thrown into question the premises on which such design awards are made. Even when their award-winning status would signify the "quality" of the architecture of the building, user evaluations of the buildings can and do challenge the existence of environmental quality as experienced by building users (Marans and Spreckelmeyer, 1982; Vischer and Cooper Marcus, 1986).

2. *Meeting Users Needs:* One exception to the building-

research-without-a-building approach is the evaluation of a government office building in Ann Arbor, Michigan (Marans and Spreckelmeyer 1981). Here, some specific design objectives for the building were identified, and data on use of the building and attitudes towards it by visitors and occupants were collected and analyzed with reference to these objectives. An interesting innovation of this approach was to identify visitors as well as workers as "users" of the building and to study the impact of the building on the local community as well as its performance as an environment for work.

The conceptual model invoked by these authors makes some fundamental assumptions about relating measurable attributes of the environment to intrinsic sources of worker satisfaction. For example, the book states that the "outcomes or indications of success in work environments" are occupants' rating of their "overall environmental satisfaction, job satisfaction, and worker performance," and that "a central purpose of evaluation research is to explore such connections between specific environmental attributes and people's perceptions of them" (ibid., 25). This study found, for example, that workers in areas where a lightwell opened above to another floor were highly likely to complain about poor visual and voice privacy, regardless of how far they were located from the lightwell. The access to daylight and window views did not, apparently, offset feelings of inadequate privacy. How these workers would feel about a more enclosed work area without the lightwell feature and therefore without window access can only be speculated upon. The connection between specific environmental attributes and people's satisfaction is often indirect.

3. *Applying User Satisfaction:* Many studies of user satisfaction in buildings are designed to elicit practical information about how best to respond to users' complaints and recommend practical building improvements that will solve building problems.

For example, R. J. Goodrich coins the term "the perceived office" to integrate the real aspects of the worker's environment with the way in which he or she perceives them (Goodrich 1982). He postulates six "interactive subsystems" in the perceived office, each of which impacts on and interacts with the others:

> 1. the kind of people who work in the office and their psychological characteristics; 2. the work, activities and tasks which these people perform; 3. the social processes, communications and relationships they have with each other; 4. the organization, its structure, style and formal and normative characteristics; 5. the type of technology that is used by people to perform their work activities; and 6. the designed environment itself (p. 358).

Goodrich reports general findings pertaining to how people in offices use their space; he does not look for a tight fit between "objective" and "subjective" correspondence in environmental assessment, or for empirical relationships between buildings and behavior that could be construed as generically true. Instead, he reports descriptive findings in specific offices that show how the designed environment is used for work, social processes, the introduction of technology, and organizational culture. This is a practical approach which has been used by managers with building responsibilities in large corporations. It does not incorporate or address the technical aspects of building performance or how to determine environmental quality at a technical level.

4. *Incorporating the Context:* Another more practical methodological approach is to look beyond the environmental needs of individual workers to the opportunities and constraints of the organization (Moleski and Lang 1982; Gray et al. 1986). One example is the "Purpose-driven Model for Building Assessment" (Zeisel 1985). In a recent office building evaluation project, the owning organization's goals and its priorities for use of the building were specified as part of the evaluation. The information that was collected on the quality of the building and how well it worked for users could then be assessed in terms of the purpose(s) of that particular building for the user organization. In this approach, the meaning of the information acquired about the building is lodged in the context of organizational priorities.

Suppose, for example, that an organization designs and builds a building to move employees from aging crowded quarters into a new, attractive, centralized facility in order to demonstrate corporate pride, organizational consolidation, and to give a sense of employee worth. Specific criteria (called *effectiveness criteria)* are defined to enable an assessment to be made of how well the building is meeting these objectives (fulfilling its purpose). Operational measures of the criteria (called *priority attributes)* direct data collection and analysis on building performance and use. And the findings feed back directly into organizational objectives and the purpose of the building.

This approach has the effect of rendering building performance problems readily understandable to the administrative decision makers who are not a priori interested in facilities. It facilitates management decisions about building change and improvement. It places building performance issues and inventory management in the context of overall organizational productivity. This is a somewhat different way of defining environmental quality. Whereas in the building-in-use, environmental quality is integral to the interaction between users and the built environment, the Purpose-driven Model does not set out to define environmental quality as such; buildings assessed using this model are considered satisfactory if they perform adequately the purpose ascribed to them by the occupying organization. Where the building-in-use attributes a psychological component to environmental quality, the Purpose-driven Model invokes the culture of the organization.

In summary, the most useful building evaluation studies are not derived from the analytic model, but from models that incorporate user-building interaction and locate this in the context of an organization. Satisfactory solutions to the three office problems of the open plan, office automation, and occupational health threats are not going to be found in user satisfaction studies, or indeed in any method of evaluation that, even unconsciously, conceptually separates the user from the environment being studied. The critical aspects of the office environment can only really be understood as intrinsic building-behavior interaction, and to be understood, should be studied as a system. A good example of something that has been studied analytically, but would benefit from an integrated methodological approach is privacy.

PRIVACY IN OFFICE DESIGN

Concerns expressed about inadequate privacy and the intrusiveness of office noise, inadequate space for work storage, and poor indoor air quality have predominated in office building evaluations that have questioned occupants on their needs, preferences, and complaints about their environment. Just as decades of housing evaluations showed us that inadequate parking and between-unit soundproofing are the generic user problems in multifamily housing, so office workers can be counted on to produce the above-mentioned list of complaints, in which lack of privacy occupies first place (Wineman 1982).

Privacy is, by definition, a cultural and situational concept. A Japanese office-environment researcher recently pointed out that the concept did not exist at all in Japanese offices until it was introduced by Western researchers and Western office planners (Ekuan 1985). Therefore, a serious problem presents itself when trying to relate people's subjective ratings (comfortable versus uncomfortable, private versus not private, satisfied versus dissatisfied) with so-called objective measures such as number and height of acoustic partitions, ambient and background noise levels, and size of enclosed floorspace area. Privacy is not amenable to the experimental research paradigm in which extraneous variables are controlled and subjects' judgments are related directly to controlled changes in environmental conditions: it is by definition context-dependent.

A number of studies have, nevertheless, shown an empirical association between subjects' ratings of improved privacy and more enclosure of the work space (e.g., Sundstrom et al. 1980; Brill et al. 1985). In these studies, worker privacy is inferred from respondents' dislike of intrusions, from reports of disturbing noise levels, and from workers' complaints about lack of control over their own accessibility by others. However, on more than one occasion, workers' attitudes were compared before and after a change from an enclosed layout to an open plan, without controlling for the impact of the change itself. And some authors admit that the data collected from office employees seem to reflect broader concerns with the quality of the office environment, and are not necessarily inherent criticisms of the design concept (Szilagyiand and Holland 1980).

To define privacy as a person's control over his or her accessibility by others, as it is often defined, is a circular argument: the more walls and doors one has, the more control one feels one has over accessibility by others, and the more privacy one can say exists, thus explaining this often-found finding. Applying environmental determinism to this argument, one finds that if lack of privacy is a major source of worker dissatisfaction, then the "design response" to lack of privacy is to enclose all office workers' desks with acoustic screens or floor-to-ceiling walls that will allow workers to control their accessibility by others. This is often what is done.

There are serious problems with this solution, however. Increasing the enclosure of the individual work station is a knee-jerk reaction that can generate more problems than it solves. First, heavy use of partitions in an open plan impedes ventilation and air flow and blocks natural light. Second, acoustic partitions provide some visual privacy but do not alone block out substantial amounts of noise. If floor-to-ceiling dividers are used, optimum office size becomes an issue (one-person offices, or two, or three?); people sit with their doors open for air and to see the office; and noise problems tend to increase.

In fact, the evidence suggests that workers' complaints about privacy in open-plan offices are universal and can be used as a vehicle to express complaints about air quality, lighting conditions, management, or boredom at work. The issue of privacy blurs the distinction between environmental satisfaction and job satisfaction on which so much office building research is predicated. One possibility is that privacy is one of those issues that users will always complain about because it is a catch-all complaint category, much as lack of storage space is in multifamily housing. Asking people if they have enough privacy is like asking them if they have enough money: some will answer, "For what?"; most people will say, "No." On the other hand, as all social programs bear witness to, giving people more money (or in this case, enclosure) does not guarantee improvement in their circumstances.

The problem posed by the privacy issue is one of experimental logic. If, for example, studies of domestic violence show that most incidents of family violence occur in the kitchen, does this mean that domestic crime would be reduced if houses were designed and built without kitchens? No, obviously not! Similarly, enclosing individuals' work stations will not alleviate their complaints about lack of privacy.

The analytic approach to building evaluation will not help find solutions to occupants' privacy problems. Typically the analytic approach to evaluation studies

> proceeds by separating out the problem in two ways: they separate out the problem of meaning from the intrinsic material nature of the artifact . . . and they separate out a human subject from an environmental object and identify the problem as one of understanding a relation between human beings and their built environment (Hillier and Hanson 1984, p. 9).

Not only does the separation of human subject from environment object obviate the definition of environmental quality developed in chapter 2, but it leaves building owners and managers with the following dilemmas:

- reconciling incompatible needs and requirements of users, and at least setting priorities on whose needs/requirements should predominate;
- understanding occupants' interaction with a building through *adaptation* of their behavior to meet the requirements of the environment, and through whatever means of environmental control they have at their disposal;
- assuring the relevance of both the *system* of evaluation used, and of the *findings* yielded by such studies to the institutional and procedural context in which the building operates.

APPLICATION OF THE BUILDING-IN-USE

The building-in-use approach is synthetic rather than analytic. It is methodologically simple enough to be used by trained and untrained researchers. Although well-grounded in the tradition of psychological research, it yields information about people and buildings that can immediately be put to use in solving building problems.

It is basic to the building-in-use approach to study a building through the experience of it by those using it, and in this way to generate useful results. It is important to remember that the purpose of environmental evaluation is to evaluate the building, not the person in it. There is a difference between using the person's experience to evaluate the building and evaluating the experience of a person who happens to be in the building. The study of privacy tends to focus us on the person's experience rather than on the building's performance; what we really want to do is to use the person's experience of the building to enable us to evaluate the building-in-use.

The building-in-use method uses occupants' judgments to refocus evaluation on the building, rather than on the user. The building-in-use is the focus of the study, because buildings without users cannot be evaluated. Through building-in-use assessment, the relative severity of open-plan office problems, electronic equipment problems, and health problems can be determined, and priorities set on solutions.

To approach building measurement through the eyes, ears, hands, nose, etc. of its users, as it were, is to measure the environmental quality, but this "measurement" must be normalized, i.e., a standard of some kind must be imposed. The methodology of the building-in-use approach applies norms and standards derived *in situ* in lieu of artificial standards derived from laboratory research. If implemented carefully, this is an easy way to assess the environmental quality of one building relative to other, similarly used buildings. The organizational context is a part of the implementation of this methodology: the environmental quality being measured is by definition contextual. The measuring system can be developed and used by a single building owner, by a corporate owner, by a group of occupants, by a union, by a building manager, or by the accommodations staff of an organization, and it can provide immediate feedback through the "building-in-use profile" yielded by the data.

The method we are calling building-in-use assessment addresses the three dilemmas of conventional building evaluation listed above. By not invoking the user-needs model, disparate user requirements do not need to be reconciled. Occupants' adaptation to the environment and the ways in which they can control the environment, i.e., the dynamic exchange between the user and the building, are incorporated into the concept of the building-in-use. The occupants themselves setting the normative standards solves the problem of what proportion of the conventional office environment is in fact supportive and satisfactory, and whether or not we are paying undue attention to users' complaints.

By being able to assess environmental quality, answers to the practical questions of office-worker complaints, how to evaluate them, and what to do about them, can be constructed. The results of a building-in-use assessment indicate when users complain, whether to make major changes to the building, to make small local changes, or not to do anything at all. They help decision makers use the knowledge that already exists on how office buildings perform for occupants, to determine if there are some generic truths, and if so, what they are. Building-in-use assessment is a way of using occupants' judgments to evaluate a building in such a way that owners, managers, and adminstrators can readily understand whether they have a good-, average-, or poor-quality building, and to decide how much money they want to spend on the environmental quality of their offices.

SUMMARY

In order to ascertain whether or not an environment needs improvement, can be improved, and has been improved, one must understand the building-user relationship. If one wants to determine what to do to improve one's buildings, one must find out more about the effects of the building environment on the people using the building. One way of doing this is to evaluate existing buildings.

There are three major problem areas in modern offices: the open plan, office automation, and possible health hazards. Although environmental evaluation is an activity with historical precedents and potentially useful results, too many studies present findings that cannot be directly applied to decisions about office environment improvement.

To carry out an environmental evaluation of an office building, using as criteria occupants' psychological needs, organizational goals, and social and management requirements, poses several weighty problems: the size and scale of data to be collected; how to organize these data; and how to analyze what they mean. Vast increases in amount of information available on specific buildings, or even on specific conditions across buildings, do not always yield greater knowledge of building quality, or how to assess it.

There are four basic reasons why a better and more inclusive method of evaluating buildings is not yet available, although building evaluation studies have accumulated useful knowledge on users' behavior in office buildings.

An analysis of the concept of privacy in offices, as defined by researchers and office-workers, emphasizes the shortcomings of contemporary office building research. Privacy is a concept that cannot be understood, evaluated, or, therefore, improved by separating actual physical building conditions from the behavior (attitudes, task performance, satisfaction, etc.) of building users. The status of research into privacy typifies to a large degree the problems of applying the analytic approach to building evaluation. To be able to use occupants' complaints about privacy as information about the environment which can be used to solve problems is the contribution of the building-in-use assessment system.

In the following chapter, three buildings are described. These buildings were selected for study to enable the quality of their performance to be diagnosed. Diagnosing building performance is an important component of building-in-use assessment. The next chapter describes how the diagnosis of building performance was effected, and how building performance diagnosis is an intrinsic step in the assessment of environmental quality.

Chapter 4

BUILDING-PERFORMANCE DIAGNOSIS: HOW TO MEASURE ENVIRONMENTAL QUALITY

Between 1982 and 1984 the Canadian government committed itself to an evaluation of several large government office buildings. The primary concern of the government agency responsible for operating these buildings was controlling energy consumption. The agency also wanted to increase the efficiency of structural performance, building systems performance, and building operation and management. They realized that inefficient energy use, materials failure, and poor indoor air quality were not discrete technical problems that could be solved through the application of routine engineering solutions. It was determined that a "transdisciplinary" approach was required in which specialists in various building science disciplines would analyze building performance and, in the case of each building, would come up with recommendations for improvement (Mill 1984). This approach to building evaluation has been dubbed *building diagnosis.*

In this chapter, diagnoses of three of these office buildings are presented in order to illustrate the diagnostic approach to building-performance evaluation. In each of these building studies, occupant judgments of the environment were integrated with the scores and measurements yielded by a panoply of instrumentation used to test interior environmental conditions in the office. In the process, a number of interesting findings emerged both about the buildings and about the ways in which people used them. A range of recommendations for building improvement came out of the diagnostic studies: some of these represented pressing issues needing immediate action, others recommended long-term changes that would improve environmental quality. The process of building-performance diagnosis is described below.

WHAT IS BUILDING DIAGNOSIS?

Use of the term *building diagnosis* has proliferated in recent years (Vonier 1983a; Miller 1985; Wilson 1984). Some organizations have gone so far as to define stages of building diagnosis methodology (e.g., National Research Council 1985). The word *diagnosis,* in a medical context, refers to the way in which a physician reads bodily symptoms and determines what is wrong with someone's health. The medical meaning of the term *diagnosis* denotes a problem: the physician assumes pathology, looks for symptoms of pathology, and invokes a diagnostic procedure to enable her to read a pattern into the symptoms. The physician uses diagnosis to determine possible causes of the pathology, on the basis of which inference she will prescribe a solution. The solution to the diagnosed problem is, in fact, a prescription: for treatment (e.g., drugs, surgery), for behavioral change (e.g., rest, stop smoking), or for doing nothing at all.

For example, an individual with a cold may have symptoms that range from a stuffed nose and coughing to headaches and aching limbs. The physician, on hearing these symptoms reported by the patient, and on examining him briefly, has a choice of diagnoses, each of which implies a different cause for the pathology. Some possible causes might be an allergic reaction, a cold, or the beginning of some more serious disease such as meningitis. Once the doctor has diagnosed the disease, as a professional she knows how to treat it. She prescribes medication, bed-rest, hospitalization, or hot lemon drinks. The important elements of the medical use of diagnosis are the following:

- it presumes pathology;
- it implies a cause for the pathology or problem;
- a prescription for solving the problem is implicit in the diagnosis.

In applying such a diagnostic process to an office building, the following process occurs. The "presenting symptoms" may be in the form of complaints from occupants: for example, they report that the air is too dry, or too stale, or it sends them to sleep. The diagnosis of the building's performance involves using testing procedures on the ventilation system and samples of indoor air that might parallel the use of physical examinations of the patient and laboratory tests by the physician. A problem might be uncovered in the air supply system: it is not delivering enough air to the interior of each floor. The "cause" of occupants' complaints might be that the variable air volume (VAV) boxes are not opening in response to the need for extra cooling in the interior parts of the building. Implicit in that diagnosis is the solution to the problem: the VAV boxes must be adjusted, and then occupants' complaints about the ventilation should cease; i.e., the symptom will disappear.

Some established ways in which one can deduce that a building is failing to perform, and that are thus possible indicators of pathology, are listed below:

- A significant number of *complaints from occupants* about some facet of the building, like the widespread complaints from people regarding upper respiratory problems, headaches, nausea, and a general feeling of ill-health that accompanied Urea Formaldehyde Foam Insulation (UFFI);
- A high rate of *worker absenteeism* in a building; occupants who are physically uncomfortable may arrive late

and leave early, they may call in sick more often, they may spend extra time on breaks or outside the building;

- A high rate of *unoccupied space* in an office building, or a rapid turnover of tenants. Assuming there are some market choices, building tenants will either move to more comfortable quarters or not move into a poor-quality building in the first place;
- A significant amount of *repair and maintenance* due to minor vandalism, breakage, or the rapid wear of poor-quality finishes and materials;
- *High operating costs,* such as fuel consumption, staff overtime, extra building management staff, costly cleaning services, etc.

These and other similar symptoms of pathology require diagnosis in order to judge whether or not the incidence of any of these symptoms is higher than the norm. The best medical diagnosticians are doctors who know how to use their knowledge and experience to determine the best interpretation of the symptoms. In contrast, the usual approach to building diagnosis is to invoke a not very extensive knowledge base about building performance and interior environment standards. Ultimately, the building diagnostician, like the doctor, must make a judgment about the cause of the pathology and prescribe a solution. The building diagnostician, like the physician, must invoke a norm of "wellness" from which the building (the patient) is deviating, and he must use his own best judgment on the nature of the diagnosis and the prescription for a cure.

In medicine, the norm of wellness that is usually applied to the process of diagnosis is the patient's own report of himself as feeling unwell, of deviating from his own norm of feeling well. In building diagnosis, the "physician," or building owner or manager, has an equivalent to the patient's self-report in the amount and intensity of building users' complaints. It is difficult, however, for the building manager to recognize when complaints from occupants are excessive enough to indicate "pain" and a real symptom of building pathology. It is difficult for him to know when building operation and maintenance costs are high enough to indicate that this building is both *different* from and *worse* than other comparable office buildings. In building diagnosis, the norm of wellness has traditionally been building standards: norms that are set outside the building's own context. These include the standards regulating the health, safety, and comfort of occupants that were discussed in chapter 2, and standards that provide general guidelines regarding energy budgets and repair and maintenance costs for large office buildings.

Chapter 2 showed that standards are incomplete and unsatisfactory indicators of building wellness—or quality—because they fail to take into account the specifics of a building's operating context, and also because they fail to take into account the experience and perceptions of the people who use and judge the building. Just as a person whose heart rate, blood pressure, and red-cell count are within standard limits of health may still feel unwell, so an office building designed to meet standards may still feel uncomfortable to its users.

USING DIAGNOSTIC TESTING FOR PREVENTIVE CARE

There are additional reasons for undertaking the diagnosis of a building's performance. These have to do with *preventive care.* In medicine, preventive care puts the doctor in the role of maintaining health rather than diagnosing illness; sickness is prevented rather than health problems being solved. Knowledge must be acquired and applied in preventive health as much or more as in treating pathology. Building managers carry out preventive maintenance on the same principle. In addition, building diagnosis for preventive care means analyzing those aspects of a building's performance that are effective, appropriate, and successful, and that could be incorporated into the design of new buildings. There are at least four significant reasons why building diagnosis should be used not just to determine pathology but to maintain "building health." They are:

1. *Alleviation of ignorance:* we need to know more about

how buildings work. It is necessary to study the outcome of building-related decisions to add to a general knowledge base about building performance that we can eventually use to make better building decisions.

2. *Increase in efficiency:* although large sums of money are spent building and operating office buildings, little is known about increasing the efficiency of those expenditures. Most professionals agree that buildings could be improved; but more needs to be known about how to spend money on improving them and how to improve the process whereby buildings are commissioned and built so that building dollars buy the best possible product.

3. *Exercise of control:* the process whereby large complex buildings get built is somewhat makeshift and is less systematic and controlled than one might expect given modern technology. Responsibility for key decisions passes through many hands, and if building failure occurs later, it is difficult to know whom to hold responsible and how the error might have been avoided. As corporate real estate holdings, in both public and private sectors, grow, and their value increases, owners and occupants of office buildings need to know more about how to control and manage their inventory. They need to exert more informed control over crucial decisions of size, shape, quality, and appearance of buildings than is presently the case.

4. *Response to impending change:* important social and technological changes are taking place that affect the role of buildings in our society. Some of the most obvious are office automation and the introduction of electronic equipment; changes in the work force, such as the rising number of women, the demand for flexitime, the work-at-home office; location changes, such as the separation of manufacturing from administrative functions; and increased user awareness of building quality, along with increasing pressure from unions for healthy and attractive work surroundings.

The definition of the building diagnostic process that best serves the twin purposes of diagnosing pathology and of preventive care is the following: *building diagnosis is the use of instruments and measuring techniques to understand how a building performs.* These instruments and techniques apply a range of building tests. Just as medical testing measures the performance of various body parts, so these tests measure dimensions of building performance such as occupant comfort, work-group efficiency, indoor air quality, thermal and acoustic conditions, illumination quantity and quality, air leakage from the building enclosure, population migration within a building, effective ventilation rates, and performance of the electrical and mechanical systems (Rush 1986). The measurements and techniques provide scores or ratings of building conditions which can be checked against, for example, ASHRAE standards for estimates of so-called building quality.

WHEN TO PERFORM BUILDING DIAGNOSIS

A typical building diagnosis might proceed as follows. A group of workers in one corner of a floor of an office building complain that they have poor ventilation, that the air is stale, and that it smells bad. In carrying out the diagnosis, building experts will study the mechanical systems. They will measure effective ventilation rates; they will trace the passage of air through the building; they will determine how low the air circulates into the space from the ceiling diffuser; and they will measure air speed, ambient temperatures, relative humidity, and the amount of carbon dioxide build-up over a working day. They may also test air samples for formaldehyde, carbon monoxide, ozone, and other known contaminants.

As there is no preset system of building diagnosis testing, the diagnosticians may have trouble knowing when they have done enough measurement to allow them to conclude that the occupants are either "right" or "wrong" about possible indoor air pollution. They may take measurements several times a day, or once a week for a month, or just once. The more they find out about the way air performs in a space, the more symptoms or likely causes of pathology they are likely to suspect. For example, the mechanical system may have to be assessed: the ductwork and sizing, the fan size and velocity, the calibration of the sensors, and so on. Another line of inquiry concerns the occupants' work experiences. For example, their level of job satisfaction, their degree of union activism, or the particular requirements of their task may all have implications for the perceived adequacy of their ventilation.

As the diagnosticians accumulate data, they will eventually have recognized enough "symptoms" to enable them to diagnose the problem. In doing so, they will state the cause of the pathology and the prescription for its solution. Thus, for example, if the sensors are incorrectly set, they might recalibrate the sensors. If insufficient air is being supplied to the space, they might increase the supply-air temperature or add more ceiling diffusers. If there are areas of stagnant water in the building, they will dry them out and make sure they do not recur. If the occupants are dissatisfied with their jobs or working conditions, they could be supplied with fans, or moved to another floor, or provided with a telephone "hot line" to the building manager. The "solution" is probably a combination of several such prescriptions. It is likely to come in the form of a retrofit recommendation that alleviates the principal symptoms of building failure; it may not solve the problem itself.

In the case of the Eastern Building, described below, giving people fans, moving them to a better ventilated area, and balancing the mechanical system might remove the symptom, but the context of the problem, or the system within which the problem is lodged, would remain unchanged. Sooner or later it becomes necessary to address the question of how often and to what extent one continues to "solve" building problems with retrofits and other expensive solutions, which do not alter the rate and intensity of occupant complaints.

Building diagnosis may also become necessary in buildings with high "churn rates" or relocation of work groups within the building. The average churn rate in North American office buildings is 30 to 40 percent per year, but it can run as high as 70 percent. When people's desks are moved, electronic equipment has to be moved, cables changed or added, walls and partitions taken down and put up, chairs and tables reoriented and relocated, furniture replaced (and sometimes there is less space for it), and circulation patterns may have to change for the passage of people and freight. With such frequent changes to office layout, careful design decisions regarding optimal desk orientation for light and sound, acoustic partitions for privacy, window access, and adequate air supply and exhaust for individuals and equipment do not routinely get made. If the building is not particularly good to start with, the effects of bad space planning and interior layout decisions are exacerbated.

Even when a building has been designed specifically for flexibility, such as the Central Building described below, problems such as glare from the overhead lighting on incorrectly placed VDT units, no air supply to interior offices where partitions extend to the ceiling, and acoustic screens blocking the entry of natural light into interior spaces tend to occur repeatedly. Each of these problems can be solved by immediate and relatively simple retrofit solutions, such as shading the luminaires over the terminals, placing one or more diffusers in the ceilings of interior offices, and removing or relocating some of the acoustic screens. Although these actions will solve the immediate problems, they will have to be repeated the next time the office moves around. This is like treating someone who is depressed with drugs alone. As soon as he stops taking the drug, he will suffer just as badly from the depression. There is a limit to the usefulness of repeating this cycle. As is noticeable in the case of the following three buildings, the principal building problems stem from a misunderstanding of building-user interaction and poor management of organizational and environmental change.

EXAMPLES OF BUILDING-PERFORMANCE DIAGNOSIS: THREE OFFICE BUILDINGS

A team of experts in acoustics, thermal comfort, illumination, thermography, electrical and mechanical engineering, indoor air pollution, and environmental psychology studied three buildings using measuring instruments, laboratory tests, interviews and questionnaire surveys, and intensive and systematic observation. The team's approach was an attempt to integrate physical (instrument) and psychological assessments of environmental quality in a field situation. Their data are presented with all their technical detail in the official reports on each building; below is a summary of salient findings (Public Works Canada 1985).

BUILDING DESCRIPTIONS

The Western Building

This is a 640,000 sq. ft., multipurpose office building completed and first occupied in 1978 (fig. 4-1). The building is roughly U shaped, and the eight stories overlook a river on one side and the downtown of a western Canadian city on the other. Between 1,700 and 1,800 people work in the building, which also contains shops, a cafeteria restaurant, and a post office. Visitors are estimated at around 2,000 visits per average working day.

The configuration of each interior floor varies slightly as a result of setbacks on the upper floors, but the floors are deep,

4-1. The Western Building.

with continuous bands of sealed glazing around the perimeter (fig. 4-2). The second and third floors enclose a two-story atrium which incorporates a government information center, a directory board and inquiry desk, an exhibition space, and access to elevators and an escalator. The primary service loading dock is on the east side of the building, with a secondary dock on the west side.

Three large mechanical systems service the building. Designed for open office layouts, air distribution, lighting, and sprinkler systems are integrated into a suspended coffered ceiling. Illumination on most floors is provided by single fluorescent tubes; heating is provided along exterior walls below the windows. There are two banks of passenger elevators in the building, and a freight elevator.

The regional offices of several government departments are located in this building. The layouts, furnishings, and types of work activity vary throughout the building, as do office size and composition. Among the types of office interiors to be found in this building are the following variations.

- open-plan office space, using acoustic partitions (screens), plants, and office furniture for separation;
- open-plan space with no barriers between work stations (known as a "bullpen" design);
- drafting tables in a large open room with task lights;
- offices with groups of eight to twelve desks enclosed by screens.

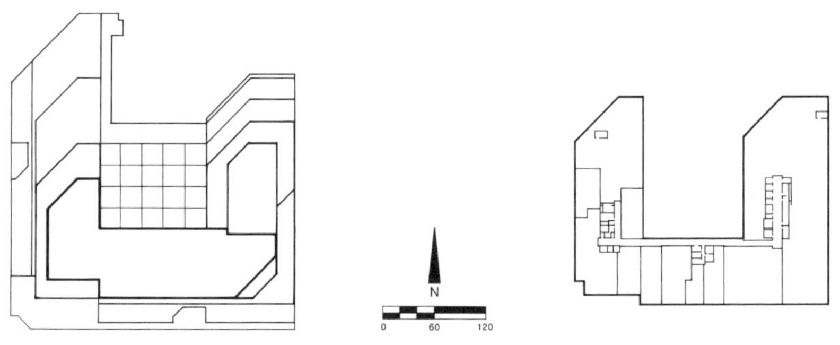

4-2. The Western Building: roofplan and a typical floorplan.

All of these types of layout include one or two enclosed offices—usually at the perimeter—for supervisors and managers. In addition to the office space, there are:

- waiting areas for the public with seating in rows, and a long counter, behind which individual work stations are surrounded by acoustic screens;
- computers enclosed in their own rooms with special temperature and humidity provisions;
- printing facilities and large machines;
- video display terminals, printers, and copying machines dispersed throughout most floors.

This variation in types of office space has to accommodate the following range of office tasks. In some offices, workers do paperwork—filing and similar clerical tasks; in others, they use sophisticated microelectronic equipment connected to computers. In some, they receive, respond to, and/or interview members of the public needing various government services. In most offices, supervision, administration, and management tasks are also performed. There is widespread use of computer equipment for data entry and analysis, and many computers are connected to printers; some workers' tasks require the reading of computer print-outs. Some workers have inspection or investigation tasks out in the community and are rarely at a desk; one agency provides fitness training in its offices; and all have varying needs for paper and equipment storage.

To expect a single office building to provide optimal conditions for such a variety of tasks and environmental requirements is unrealistic unless special steps are taken. The building interior had not been programmed for specific space uses and work-group layouts; each agency or organization had submitted a list of the number of workers with space and equipment requirements. Details regarding work-group size and relationships, future equipment needs, and special task considerations (e.g., acoustic privacy for employee counseling, task lighting for reading computer print-outs) had not been part of the space programming or interior design processes, with the result that the wide variety of task and work-group organization was essentially accommodated in unvarying, standardized spatial conditions throughout the building.

In spite of what might be considered a recipe for a "poor fit" situation, the building was not generally disliked by the people who worked there. In fact, both their ratings and the instrument measurements yielded an overall assessment of the building as being an average-to-good-quality modern office environment.

The Central Building

This is a 30,000-square-meter, two-story office building located on a suburban site some seven kilometers from the downtown of another western Canadian city (figs. 4-3, 4-4). The long (219 meters), rectangular, two-story building comprises two vast interior open floor areas and was designed to accommodate a specific set of activities for one federal government department. It was occupied in 1979.

The permanent population of workers in this building is around 900, but another 600 to 800 persons work temporarily in

4-3. The Central Building.

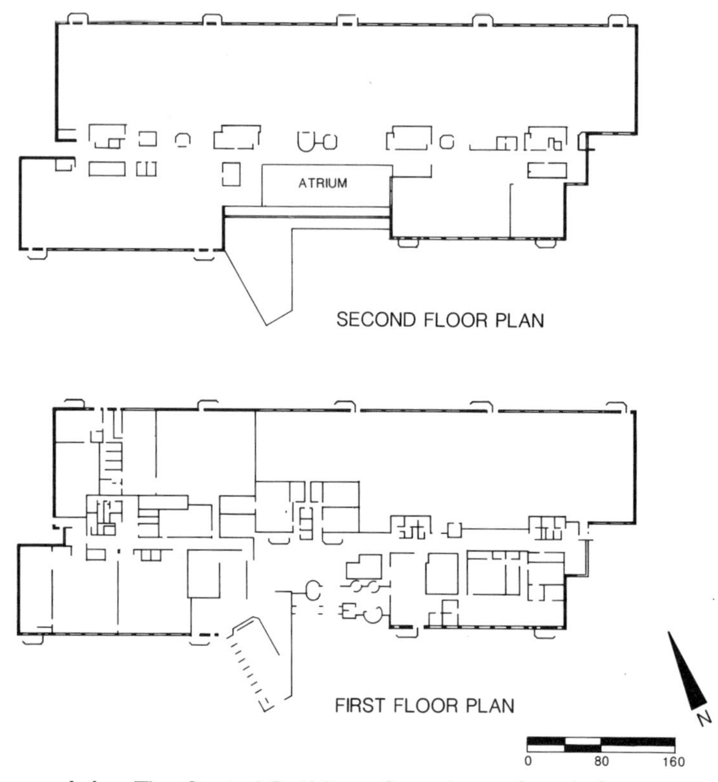

4-4. The Central Building: floorplans of each floor.

the building between March and July every year. There are no visitors as the work performed in this building is confidential, and the building is secure with restricted access. The building contains a cafeteria with an outdoor patio, and a small shop; it includes a two-story atrium in the front and a loading dock in the rear of the building.

The building is serviced by some fifteen separate air handling systems, with air supplied through ceiling diffusers and returned through the ceiling plenum. There are extensive bands of sealed windows along the north and south (long) façades on both floors, and artificial illumination is supplied through single-bulb fluorescent fixtures in a coffered ceiling.

Unlike the Western Building, the Central Building was programmed and designed to accommodate a specific set of tasks for workers whose organizational and demographic characteristics were known to the designers. The Central Building was designed specifically for the purpose of processing individual tax returns, the vast majority of which arrive between early March and the end of April each year. The "assembly line" of paper-processing activities that comprise tax-form processing are collectively known as "the pipeline," and they extend from the moment of mail delivery to the final filing away of an assessed return. Most of the short-term temporary employees are hired for tasks on or related to the pipeline, whereas the permanent workers perform other related activities for the income tax department.

The overall purpose of the building is to process paper. Several million tax returns are processed through the pipeline every year. Unlike the tasks carried out in the Western Building, the tasks of Central Building workers are not vastly different from one another. Work groups vary in size, but the second floor is virtually completely open, with few space dividers of any kind. People work at desks sorting, reading, classifying, and organizing mounds of paper; large areas of the floor are dominated by data entry and data processing tasks for which large banks of VDTs and other types of electronic equipment are used. On the first floor, the more conventional office tasks related to the ongoing operation of the organization are performed.

Whereas pipeline workers do not require privacy, interpersonal communication, or long-term document storage in the work area, the other, more conventional, office tasks do have such requirements. However, almost all work groups have an open-plan office layout. The paper processing tasks of the pipeline require adequate and sensitive lighting, adequate cooling around the electronic equipment, and enough flexibility in building design to function as effectively and efficiently for 2,000 workers as for 800. The building was judged generally adequate by its occupants, who were surveyed twice. On the first occasion, there were approximately 1,000 workers in the building, most of whom were permanent. On the second occasion, when the pipeline was in operation, there were some 1,700 workers. In general, the second group had a more positive attitude regarding the quality of the work environment than the permanent group.

The Eastern Building

The Eastern Building is a multi-tower highrise building located near a large eastern Canadian city. The building is very large (it accommodates some 3,000 workers) and is in the form of three connected towers ranging in height from eight to twelve stories (figs. 4-5, 4-6). Not all the towers were tested. Two floors in one tower were sampled, and four floors in another. The first two floors (2 and 3) have deep interiors and a long, nonrectangular perimeter with a continuous band of windows. They are in a twelve-story tower served by two large mechanical systems, and air is supplied by a VAV system through diffusers at the perimeter, and returned through ceiling vents to a duct. The ceiling system is coffered, as in the other buildings, with single-lamp fluorescent fixtures overhead and no task lighting. Both floors have conventional open-plan layouts that use 60- and 65-inch acoustic screens to enclose individual work stations and small work groups. Circulation through the floors is directed by "walls" of 72-inch screens.

4-5. The Eastern Building.

The other four floors (4,5,6, and 7) are in another tower, eight stories high, whose floors are rectangular and less deep. The tower is served by one large mechanical system that supplies air through ceiling diffusers and to which it is returned via the ceiling plenum. The fluorescent lighting on these floors

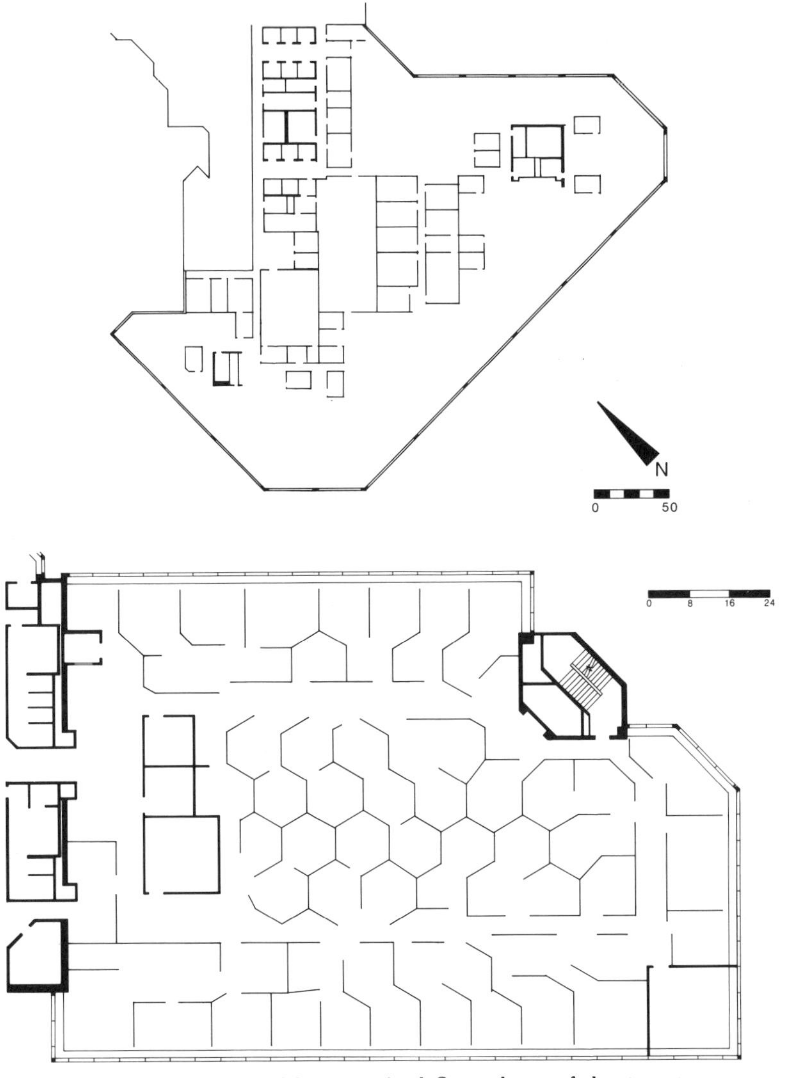

4-6. The Eastern Building: typical floorplans of the two towers.

is not recessed in ceiling coffers but is supplied from single-lamp parabolic fixtures with egg-crate diffusers. A randomized-octave ("pink noise") sound-masking system has been installed. Although these floors are smaller, they are more densely used, with 72-inch acoustic screens surrounding almost every work station and less room for circulation.

Although the two towers are occupied by different federal agencies, the types of tasks performed in them are comparable and more typically officelike than in either of the other two buildings. Professional/technical- and managerial-level staff are accommodated in both towers, as well as clerical employees. All the floors have a number of VDTs, copiers, printers, and other electronic equipment scattered throughout the work areas. These are not used as permanent work stations, as they are in the Central Building, but are shared items for use by a work group or team. They are, for the most part, placed with little regard to heating, lighting, and acoustic requirements. In addition to individual work space, all the floors have staff support areas, such as meeting and conference rooms, washrooms, and staff lounge/lunchrooms.

The Eastern Building has a large atrium with connections to shopping malls, restaurants, and other large office buildings. In general, occupants like these amenities but rate the building overall more negatively than the Western or Central buildings. The building is served by a large underground garage, but occupants report parking and transportation problems among their most serious complaints about the building.

NOTES ON THE DATA COLLECTION PROCEDURES

People's judgments of environmental conditions, such as lighting and temperature, of the layout of furniture and work-group circulation, and of the value of amenities such as the shops and the atrium, were elicited through a detailed questionnaire given to a randomly drawn sample of workers in each building. Table 4-1 shows the number of respondents in each building.

Occupants were asked to assign a judgment on a scale of 1 to 5 of their perception of specific environmental conditions over the length of time they had worked in the building: these

TABLE 4-1
BUILDING POPULATIONS AND SURVEY RESPONDENTS

Building	Population	Sample Size
Western Building	1,800	400
Central Building		
(first survey)	1,000	700
(second survey)	1,800	800
Eastern Building		
(Tower 1)	550	250
(Tower 2)	800	300
Total Number of Respondents		2,450

ratings or judgments of acoustics, illumination, thermal comfort, and air quality are longitudinal or trend-monitored experiences telescoped into one perceptual judgment to which a number or rating is assigned. Occupants also answered questions related to the nature of their work, their location in the building, and the amount of control they had over environmental conditions (such as temperature) in their work space.

While building occupants were responding to the survey, a variety of individual work stations, along with support spaces such as meeting-rooms, corridors, and underground garages, were tested with instruments measuring ambient environmental conditions. These included, for acoustics, background noise levels under occupied and unoccupied conditions, and sound reverberation and transmission rates, from which noise criterion curves and articulation indices could be plotted. For air quality, air samples were analyzed for heavy molecule chemicals using gas chromatography, and for light molecule chemicals using mass spectrometry. Carbon monoxide, formaldehyde, and carbon dioxide levels were monitored, and particulate levels were measured. For thermal comfort testing, air speed, dry and wet bulb measurements of relative humidity, ambient temperatures, mean radiant temperature, and clothing and activity levels were collected to predict whether a space fell within thermal comfort curves. Air circulation patterns were studied using helium balloons and smoke pencils. The pressure readings of each mechanical system were taken, in the ducts and at the point of delivery of air to the work space, and the temperature, humidity, and volume of air delivered to interior spaces were compared to the readings printed out by the computerized air handling control system (Public Works Canada 1985).

These tests were not directed at any particular problem but were designed as a comprehensive analysis of office building performance. These buildings were considered normal, rather than pathological, examples of the government's office building inventory.

RESULTS OF THE DIAGNOSES

The Western Building

As pointed out in chapter 2, one of the complex aspects of evaluating environmental quality is determining who the users of an environment are. In the case of the Western Building, there were at least two important groups of users: the government workers whose offices were in the buildings, and the large number of visitors and members of the public they received every day. For visitors, the building was less than satisfactory. Air quality testing in waiting rooms indicated that inadequate amounts of air were supplied when these rooms were fully used. The air became stale and hot, and this caused discomfort both to the visitors themselves and to the people working in the associated offices.

Visitors were tracked and observed as they moved through the building. The behavior of visitors provided many cues as to the inadequacy of the signage. Pieces of paper taped to doors and walls on which handwritten messages redirected visitors were indications that signage and wayfinding was a problem. It was determined that building commissioners received an average of 275 inquiries in half a day.

The building performed better for those who worked in it. Lighting measurements found that lighting quality had deteriorated over the life of the building, providing workers with less light at their desks than had originally been specified. However, there were few complaints about lighting in this building,

although in certain areas where workers performed close visual tasks such as drafting, desk lamps had been introduced. Perhaps the overall reduction in footcandles to an average of 50–70 from the overhead light fixtures was more to workers' tastes than the original 90 footcandles specified in the design. Even so, 17 percent of the 400 respondents rated the lighting as uncomfortably bright.

Occupants reported feeling overheated in the building, but the thermal comfort measurements did not show major deviation from ASHRAE standards. As the ventilation appeared adequate, it is unlikely that occupants were reacting to stale or dead air, the presence of which is often reported as unwanted warmth. They may have been reacting to the dryness of the air, whose relative humidity (r.h.) was measured at various times and in various places at 20 to 25 percent. An examination of the air handling systems showed that the humidifiers were not working, and at the time measurements were taken, the outside air was also dry.

The problems most likely to be reported by occupants of the Western Building were noise and poor acoustic privacy. Instrument measurements of sound levels and noise transmission and reverberation failed to yield an obvious explanation for this finding in terms of excessive noise in the offices. However, the building accommodates a wide range of office tasks in mostly open-plan arrangements, and the acoustic discomfort reported by occupants indicates inappropriate "fit" between work-group layout and actual task requirements. For example, people who need voice privacy because their job involves employee counseling or confidential telephone conversations are not likely to perform well in an open plan, even if their desks are surrounded by acoustic screens. The measurements did indicate an unusually low background noise level in this building. The mechanical systems, though large, had been well baffled, and office workers who may have relied on noise from the air system to help mask private conversations were exposed to a more silent environment than suited them.

The next most frequently reported complaints from Western Building workers were lack of fresh air and poor ventilation. Although the air handling systems were sized more than adequately to serve the building, the perceived lack of fresh air ("staleness") may have resulted from poor air circulation in the work spaces, from the recirculation of smoke from the 42 percent of occupants who smoked cigarettes or pipes, and from the low relative humidity.

Further analysis showed that the spatial layout of the work group and individual furniture of the work station were considered more important to occupants' comfort than ambient environmental conditions such as heat, cold, light, and acoustics. Whereas 40 percent of respondents felt satisfied with their work-station furniture and with the layout of the group work space, only 10 percent were satisfied with their voice privacy, 13 percent with the noise level, 20 percent with the ventilation, and 21 percent with the ambient temperature. Given these figures, it is fortunate that the feature that satisfied most people was also the most important aspect of environmental comfort. However, this was a building with generally high satisfaction levels, no known building-related sickness, and no history of occupant complaints. In a less successful office building these priorities may not hold.

The Central Building

In the Central Building, the questionnaire was administered on two separate occasions. The thrust of the analysis was to compare the building under half-full conditions (typical for most of the year) with its performance when fully populated (from March through July). Unlike the Western Building, there were no visitors to this building, but there were two groups of office workers: the permanent workers who worked there year-round, and the temporary workers who came for all or part of the pipeline period between early March and early July. Both of these groups are legitimate building users whose experiences can be compared.

The permanent full-time and long-term workers were more likely than the temporary occupants to assign poor ratings to their work environment, both at the microscale of the individual work station, and at the macroscale of the entire building. For example, 39 percent of permanent respondents and 31 percent of temporary respondents thought they had poor daylighting; 53 percent of the former and 43 percent of the latter found the temperature shifts uncomfortable. Moreover, 56 percent of permanent occupants and 47 percent of temporary workers were critical of the ventilation; and 55 percent of the former and 45 percent of the latter complained of poor voice privacy. This pattern of response suggests that workers became more critical and adapted less well to adverse or uncomfortable environmental conditions over time, whereas the short-term workers had no long-term investment in the quality of the environment and were satisfied with a well-paid opportunity for temporary employment. Measurements of air quality and ventilation did not show any significant differences between the full and half-full operating conditions of the building. In both conditions, the building's air handling systems met air quality and ventilation standards.

The conditions rated most negatively by both populations were related to indoor air quality: all users complained of dryness and stale air. In areas where large numbers of VDTs had been installed, the cooling capacity of the chillers was found to be inadequate during peak operation in early summer, even though the design rationale for the relatively high number of mechanical zones in this building was to provide a building interior that was flexible enough to accommodate major population and task changes on a cyclical basis.

The thermal comfort measurements found inadequate humidification (r.h. was measured at 25–30 percent) and inadequate temperature control resulting from poor location of thermostats. Users indicated by their ratings that frequent temperature shifts were a problem and that they were often uncomfortably cold (nearly 50 percent of respondents).

Improved humidification may have alleviated the effects of the large amounts of paper dust that was generated by the pipeline. Testing for respirable particulates did not identify excessive paper dust in the air, but the filters in the air handling systems became very clogged when the pipeline was in operation. Workers reported sore eyes, sore throats, and some skin irritation, which could have been reactions to excessive fine dust in the air.

The paper processing tasks that were accommodated in

this building placed significant visual demands on workers. The paper was multicolored, and the print varied both in color and size. The lighting in this building was not tailored to task requirements. Uniform high levels of overhead fluorescent lights resulted in glare and reflections at the work surface and also caused glare on terminal screens. Additional lighting measurements showed that for many of those working with VDTs, the white walls and large windows of the visual background were also a source of glare: the luminance ratio of screen to surround ranged from 1 : 3 (the desirable ratio) to 1 : 12, 13, 15, and even 17. A large percentage of occupants in this building—around 60 percent of the building population—reported some eyestrain.

Work groups who were not part of the pipeline and had conventional office needs, such as screening for visual and acoustic privacy, space for paper and file storage, and special ventilation for equipment such as copiers, microfiche readers, and personal computers, suffered from a lack of conventional office space. The large open-plan floors, while accommodating the rapid transfer and movement of documents in the pipeline, were ill-suited to office workers using telephones and filing cabinets. Some of their needs could have been met by installing systems furniture and breaking their office space up into manageable work-group size, but as it was not known how much productivity was lost from the adverse effects of the wrong type of office space on this small (but essential) work group, no improvements had been made.

Many of the occupants' concerns were alleviated by the attractive cafeteria, the light airy atrium, and the attractive outdoor appearance of the building. The designers had counted on the overall attractiveness of the building being an incentive to people to work there, and it was.

The Eastern Building*

The findings from the Eastern Building illustrate more fully than those from the other two buildings the importance of occupants' psychological viewpoints in diagnosing building performance. In this building, environmental changes had been effected on

* Only one set of data from the Eastern Building is discussed here.

one of the floors in response to a specific set of complaints from building occupants regarding ventilation, air quality, acoustics, and lighting. Workers' judgments of their environmental conditions were elicited from occupants of this floor and from another floor of the same tower which was comparable in most important respects, but in which no major changes had been implemented since the building was first occupied.

Owing to the fact that one office floor had been improved, it was expected that the responses of occupants of the changed floor would show more favorable judgments of environmental conditions than responses from the unchanged floor. The full battery of building performance tests was not carried out in this building. Instruments were used to measure effective ventilation and indoor air quality on both floors by testing temperature and relative humidity, particulate levels, CO_2 levels and level of formaldehyde, and by analyzing air-flow patterns (Public Works Canada 1986).

Improvements to the changed floor included fewer acoustic partitions (to improve air flow), more acoustic baffling of hard, sound-reflective surfaces, more desks oriented to the windows for daylight and views, rebalancing of the air handling system and addition of some air supply and return vents, less electronic equipment, additional (overhead) lighting, and more colors on the walls and furniture.

Surprisingly, those environmental conditions disliked most by building occupants were rated equally negatively by occupants of both floors—in other words, changes to one of the floors did not make a difference to occupants' perceptions of these conditions. Some of these results are listed in Table 4-2. The table shows that most conditions were rated *worse* by the improved floor than by the unchanged floor. Relatively few items were rated better by changed-floor occupants. In fact, voice privacy and ventilation are strongly disliked by users of this building and the improvements to one floor did not alter their attitudes. The better ratings for daylighting on the improved floor suggest that the changed furniture layout, with lower acoustic screens, more desks orientated to windows, and less perimeter circulation space, did in fact improve occupants' experience of windows and daylight.

The results of the air-quality analyses showed *no significant*

TABLE 4-2
DIFFERENCES IN OCCUPANT JUDGMENTS BETWEEN THE IMPROVED AND UNIMPROVED FLOORS*

Complaint	Changed Floor (%)	Typical Floor (%)	Total Complaints (%)
Poor ventilation	no difference		68
Air too dry	40	25	33
Uncomfortable lighting	20	8	37
Inadequate daylighting	30	38	57
Uncomfortable noise from air-handling systems	26	11	19
Poor speech privacy	no difference		79

* Not all the ratings received from occupants of environmental conditions are reported in this table.

differences in indoor air quality between the two floors. Air circulation and carbon dioxide tests showed that the ventilation system of the improved floor was performing measurably better than that on the other floor. Occupants of both floors complained about the dry air, and the tests showed the r.h. to be 23 to 25 percent on both floors. How can this pattern of user assessment of the environment be explained?

The negative attitude of workers on the changed floor can be attributed to two factors. One is the predominance of female employees on the changed floor. Statistical analysis showed that the "femaleness" of the population accounted for much of the negativity in judging environmental quality. Women in offices have been shown to be more sensitive than men to ambient environmental conditions such as temperature and ventilation, and more prone to complain about headaches, eyestrain, fatigue, colds and coughs, muscle pain, and general stress (Rohles and Jones 1983; Nelson et al 1984; Brill et al 1985; Vischer 1987).

Another possible explanation is that when the building was first tested in response to occupant complaints, real problems were found in the operation of the air handling systems and the indoor air quality. However, like the workers in the Terrasses de la Chaudiere, the Eastern Building occupants had to wait to be heard, for the building to be tested, and for changes to be made. As a result, they are now "sensitized" to the problems of their building and are therefore almost impossible to satisfy.

In the case of the improved floor, little more can be done to "improve" the environment for the workers. The money budgeted for improvements has been spent, and objectively the floor has been improved, though with little positive effect in terms of increased user satisfaction. However, the building managers have a right to know whether or not it would be worth investing in similar improvements to the other floors of the building. The anticipated payoff from greater occupant satisfaction or productivity is clearly not the criterion by which to make such a decision. On the basis of these data, managers could "improve" the other floors in the building by removing female workers! In spite of building improvements, the demographic and political issues that are real to office workers have resulted in quite a different assessment of environmental quality. The difference between environmental *change* and environmental *improvement* is the difference between reacting to user complaints and making planning decisions based on the building-in-use.

BUILDING DIAGNOSIS AND THE BUILDING-IN-USE

The findings from these three diagnostic studies of admittedly normal buildings are not shocking: they reveal some good points and some problem areas but nothing that might warrant immediate, dramatic changes to the current state-of-the-art office building to rectify dire conditions of worker ill-health or profound discomfort. They show that most of the people do all right most of the time. The recommendations that emerged out of these diagnoses enabled the building staffs to rectify the "pathology," and for the most part generated a modicum of improvement to environmental quality in these buildings.

However, the diagnosis does not tell the whole story of building performance. Just as the Central Building is a more satisfactory building to its temporary, short-term population of users than it is to its permanent workers, so the Eastern Building is more of a problem to female than to male employees. This is a reality that affects whatever types of change and improvement managers effect to the office environment. Complaints about ventilation may mean anything from real contamination of the air supply to an incident in the building's history, or just general fear and anxiety in the user population. Complaints about noise may mean anything from loud intrusive noises caused by equipment (too much noise) to insufficient background noise to mask speech (too little noise). Complaints about privacy, as pointed out in chapter 3, can mean anything at all: low pay, unpleasant coworkers, bad lighting, new manager, uncomfortable furniture, as well as the legitimate inability to control one's accessibility by others.

In the Central and Eastern buildings, demographic characteristics affected users' judgments of environmental quality; in the Western Building, many of the users' negative judgments were caused by the demands of their tasks. As a result, we are not in a position to understand the *relative* meaning of users' judgments. Users' complaints are not necessarily what they appear to be: to what should people's environmental ratings be compared? It is important for decision makers to know what a normal rating or occupant response would be in order to understand the meaning of occupants' complaints.

The diagnostic approach provides comprehensive information on building performance, but there is nothing to compare it to. It is possible, for example, that although the judgments of workers on the two floors of the Eastern Building deviated from one another in an unexpected direction, when taken in a larger context of building wellness, they show that the Eastern Building is "normal" with regard to its ventilation characteristics and indoor air quality. The definition of wellness for these buildings is not explicit: the environment met standards, but people reported pathology. Did the diagnosis really discover pathology in the sense of a state of affairs that deviated from wellness? Occupant responses from the other tower of the Eastern Building suggest that a degree of discomfort with air quality and ventilation *is* normal in these buildings, and other studies which provide knowledge of behavior in office buildings (see chapter 3) indicate that complaints about voice privacy are normal in any open office, with or without acoustic screens.

The goal of the building-in-use approach is to solve the problem of understanding the meaning of users' environmental judgments by incorporating *user expectations* or *user norms* into the definition of the environmental problem. User norms of wellness are likely to be more useful as a basis for making decisions about environmental change than ASHRAE or other standards of building quality.

SUMMARY

Building diagnosis is the application of a battery of tests to a building to analyze and assess the quality of its performance. Three Canadian government office buildings were diagnosed, and an overview of the findings shows that occupants are particularly concerned about ventilation and indoor air quality, and that groups performing a range of office tasks require more acoustic privacy and better noise conditions than workers involved in assembly-line paper processing.

Some significant patterns of user judgments of environmental conditions relate to social and demographic factors such as the sex of the respondents and whether workers are permanent or temporary. As a result, it becomes difficult to assess whether or not changes to the physical environment in response to workers' judgments represent improvement. It is important to understand the user-building relationship as a system in order to effect *improvement* rather than just *change*. Unless managers understand users' perceptions and recognize that their judgments of environmental conditions are as much an indicator of these perceptions as they are of "real" environmental problems, they will never be able to assess the effectiveness of their environmental improvement activities.

In terms of preventive care, the net effect of these findings is more complex. The ways in which user groups are defined is a crucial component of how a building performs. Physical building improvements do not alone ensure improvements in environmental quality, and the psychological conditions of users are an important determinant of environmental quality. The building-in-use approach uses building-performance diagnosis techniques to provide a simpler and more effective way to diagnose office building performance, a way that will not only make allowances for but will in fact *incorporate* the constraints and opportunities of social, demographic, and temporal differences. This approach uses the norms created by the situation itself, by the building-in-use, to judge the level of environmental quality present.

In the next chapter the data from these three building studies are taken a step further and are analyzed more rigorously to generate user norms of environmental quality. The analysis will show that building occupants do not experience the building in the same ways that building performance is traditionally measured. This is one reason why building performance diagnosis is not able to provide the complete answer for decision makers seeking to solve building problems.

Chapter 5

AN ALTERNATIVE APPROACH TO ENVIRONMENTAL EVALUATION

The problems in the three buildings described in the previous chapter are easily recognizable by building managers and are not difficult to solve. When managers receive enough complaints (symptoms of pathology), they carry out tests and measurements of environmental conditions (diagnosis) and eventually effect retrofit improvements to fix the building's problems (cure).

Although standards for health, safety, and comfort can be applied to assist in the diagnosis, some building problems exist for which there are no solutions. For example, occupants in an office building may complain about an unpleasant odor. The odor has to be prevalent or to occur at the right time for managers or their experts to smell it too. Tests may be carried out on the air handling system, the plumbing, pollution migration, and outdoor air movement; but if nothing untoward is discovered that can explain the noxious odor, the only indication of pathology is the users' complaints. Every organization needs to take a stance on how to respond to building users' complaints; there may be no "objective" way of verifying them, but managers ignore them at their peril.

INCENTIVES FOR BUILDING IMPROVEMENT

A lack of corporate policies regarding the environmental quality of office buildings traps building managers into a *reactive* role. If workers are ill, or physically endangered, or complaining about being environmentally uncomfortable in their offices, then the owning or operating corporation can probably be persuaded—on the grounds that uncomfortable or sick workers are not productive—to spend its building dollars on fixing the problem. A corporation can do this without making a long-term commitment to environmental improvement. Why no long-term commitment? Because there are no clear yardsticks for environmental quality. As a result, should managers desire to adopt a *proactive* role and advocate for better environmental quality and the possibility of improvement, they are likely to be frustrated. Convincing senior management that it is worth investing money in environmental improvement with a long-range planning perspective and commitment of resources because it is ipso facto a good thing is difficult, particularly when pitted against forceful arguments based on the dollar returns and calculable payoffs from other parts of the organization.

Most corporate (public and private) building owners will, however, agree that they want the quality of their building stock to be good rather than bad. It is quite difficult, however, to determine what "good" and "bad" really mean in this context. The previous chapter pointed out that diagnosing building performance can indicate problems and successes but will not alone show whether an office is better or worse than the norm or is an average building. If there is no way of answering this question in "hard currency" terms that are directly meaningful to senior management, then the usual rationale for evaluating a building—i.e., that in some nonspecific way knowledge about buildings will accumulate and environmental designs will get better—is about as forceful as saying that if we all gave up our cars and used public transportation, there would be less air pollution and we would have healthier surroundings.

As a general social goal, who can disagree with this argument? The only way to get people to do it, however, would be to specify the number of cars to be given up, to provide target figures for the reduction of pollutants in the air, and to be able to prove to those giving up their cars that their health was improving—for example, that they were living longer. In other words, one would have to place manageable and comprehensible limitations on the problem statement and define specific outcomes. In this chapter we show how the building-in-use approach can supply this kind of strategy to the management of office buildings, so that each specific scale and type of building improvement will result in a decrease in the number of complaints.

STRATEGIES OF ASSESSMENT-SYSTEM IMPLEMENTATION

Building-in-use assessment provides a basis for answering the question of whether or not the building or building stock in which one is interested is of good, medium, or poor quality. It provides a basis for comparing the results from one diagnosis of building performance to results from other buildings.

The basis for a building-in-use assessment of environmental quality is a comparison of buildings or parts of buildings to one other. The parameters of this assessment are set by the agency performing the assessment, so organizational values and goals are incorporated into the approach. For example, in the case of the federal government, the same kinds of tasks and work organization exist in all their office buildings, which accommodate federal agencies all over the country. Thus, the performance of one building can be assessed against the performance of the rest of the inventory, or a sample thereof, regardless of location. By the same token, a government agency in Canada would probably *not* want to compare the quality of one of its buildings to that of a corporate headquarters office tower in Southern California.

In the building-in-use approach, environmental quality is not considered to be absolute but is seen as relative to its context. This context can be defined in many ways, depending on the needs of the organization or person initiating the assessment. The referent buildings that comprise the context may be owned by the same organization or occupied by other divisions of the same organization; they may be occupied by different groups all of whom perform similar tasks, or located in the same part of the country, or built by the same architects and engineers, or a combination of the above. The basis for their commonality is postulated by the person who wants to know which ones are good, or whether one or more are particularly bad. It is also useful to compare parts of buildings: floors of a building can be compared to one another, southwest corners can be compared for every floor, or offices with electronic equipment can be compared to conventional office interiors.

What are the most propitious opportunities to implementing building-in-use evaluation? It can be used, for example, in buildings where a few people are complaining about some aspect of building quality, and previous attempts to solve the problem, by, say, purchasing a new furniture system, or rebalancing the air handling system, or giving everyone desk lamps, have not reduced occupants' complaints. A building-in-use assessment would yield a more precise idea of what is wrong and what to do about it.

Building-in-use assessment need not be generated by occupant complaints, however. Environmental quality can be assessed at any time and for any number of reasons. Perhaps the most useful of these is to provide a rationale and a strategy for what is usually called *preventive maintenance.* The results of a building-in-use assessment will show which aspects of building performance work best for users and which are sources of dissatisfaction. As these priorities are set by users themselves, the program of maintenance and renovation that addresses these priorities is by definition highly responsive to building occupants. Managers may also implement a building-in-use assessment to help them develop accommodation standards, to trouble-shoot a "problem" building, or to take stock of the quality of the inventory.

The point about the building-in-use assessment system is that there is a built-in balance between the good and the bad aspects of a building's environment. Users of the building-in-use assessment system will be able to recognize aspects of building excellence and to tag building performance problems. Comparing a building to other similar buildings has the strategic advantage of offering a manager a way out of the dilemma of improving (or at least changing) buildings to reduce occupants' complaints (reactive) versus trying to improve buildings for the social good (proactive).

USING RATINGS TO ASSESS BUILDINGS

The organization wishing to implement this system begins by setting up a database that provides a baseline of ratings to which all buildings or parts of buildings may be compared. This is done by initiating a series of occupant surveys in which the appropriate rating scales are tested, normalized, and then used as standards. Some technical expertise may be required for scale design and testing, but once the database is established, it can be used and reused indefinitely with intermittent updating.

What is an appropriate rating scale? Such scales are designed to test the quality of ambient environmental conditions in the office. Users are typically asked to assign a rating of 1 (uncomfortable) through 5 (comfortable) to various interior environmental conditions that pertain to lighting quality, thermal comfort, ventilation and air quality, furniture layout and physical comfort, natural light and colors, and various acoustic conditions.* Respondents rate these conditions for their individual work station, in their work group, and on their floor. Because people from all over the building answer the same questions, ratings can be averaged for individual floors, parts of floors, or for the whole building. As each scale extends from a low score of 1 to a high score of 5, a score for each environmen-

* There are other ways of designing rating scales. The statistical rationale for selecting this type of scale is available in a technical report (Dillon and Vischer 1988).

tal condition can be represented by the average of all scores for the relevant scales in that space (room, floor, building). This single score can, obviously, range from 1 to 5, but is most likely to be around 3 (representing a normal distribution of responses) unless there is something about that particular environmental condition that is unusual—either unusually good or unusually poor.

From this score or averaged rating, the facilities staff has two useful pieces of information. First, they know whether an interior environmental condition in one of their buildings, such as ventilation, is a 2 (poor), a 4 (good), or a 3 (average). Even without normative scores to compare it to, it is useful to have a single numeric indicator of the quality as judged by occupants of various environmental conditions. If ventilation rates a 2, and lighting rates a 4, then managers have instant feedback on their building occupants' priorities and concerns.

The second piece of useful information that management has from this rating is whether or not specific spaces in buildings, or specific buildings in groups, are better or worse than the norm for the particular universe (e.g., all the buildings, or all the floors) that constitutes the basis for comparison and that provides the "normative score" for the environmental conditions.

Once the baseline of normative scores has been established, decision makers use the scores on various conditions to compare individual buildings with the norm to ascertain whether or not there is significant deviation. For example, suppose we have some doubts about one particular office building, about which we have received some complaints concerning indoor temperatures. We get environmental ratings from occupants of this building and others like it to which we would like to compare it, and we discover that the average rating (or normative score) for thermal comfort across all the buildings is 2.5. When we look at the thermal comfort rating for the individual building that concerns us, it turns out to be 2.4. This is so close to 2.5 that we can conclude that thermal comfort is not the problem in this building (that is to say, any more than it is in the other buildings!). This does not mean that there is no problem but rather that changing temperatures in this building will not solve it. The scales on which the thermal comfort score is based have been specially constructed and tested to gauge the experience of temperature by office occupants. The score on a global environmental condition such as temperature represents the sum of scores on several rating scales. The exact nature of the rating scales used is discussed in chapter 6.

On the basis of the relatively small investment required in time and money to collect occupants' ratings, to establish the database, and to compute norms for the environmental conditions in which they are interested, managers can apply the building-in-use assessment to several useful planning purposes. They can

- *pinpoint* buildings needing urgent or early attention (i.e., those that have the lowest scores);
- *develop goals and objectives* for long-term planning and resource allocation for building design, construction, and maintenance;
- *assess and monitor* the quality of any part of any building by referring individual scores to the organizational norms;
- *anticipate* and possibly control the impact of a building change or redesign of the interior environment.

AMBIENT ENVIRONMENTAL CONDITIONS

Let us now proceed to the detailed questions of which ambient environmental conditions should be rated for the database. In the diagnostic approach to building assessment, the building conditions that are usually studied fall into generally recognized categories or disciplines of building performance. These categories have evolved according to available instrumentation and techniques of measuring occupant comfort; they are thermal comfort, acoustics, visual comfort (or lighting), and indoor air quality. More recently, the topic of ergonomic comfort and human factors in the design of furniture and office layout has developed into a building performance category (Lueder 1986).

For each diagnostic category, a history of measurement exists from which standards of human comfort have been and are being developed for application to indoor environments. Much of this measurement has been carried out in laboratory rather than field (i.e., building) settings and has been constrained by available instrument technology. More sophisticated technology has enabled more precise and quantifiable definitions of the environmental conditions being measured. For example, "thermal comfort meters" are now used instead of thermometers to measure several aspects of the thermal environment and predict the degree of human thermal comfort.

Each of the diagnostic categories comprises its bundle of measurable physical attributes, which are subject to a series of computations to render them meaningful in terms of human experience. The most developed of these categories is thermal comfort. The acoustic package of measurements and computations is less well-integrated. The least developed is probably indoor air quality measurement, although this is a rapidly evolving discipline. In the following pages, each diagnostic building performance category will be briefly explained.

THERMAL COMFORT

Historically, thermal comfort does not refer to the quality of building interiors. It is a term that describes the experience of a *person* under a variety of thermal conditions. It certainly means more than simply measuring indoor air temperature: "A major goal of research on temperature has been the development of a single numerical index of thermal comfort" (Sundstrom 1986). Although scores on this index refer to the degree of personal comfort ascribed to the individual, the index score is arrived at through instrument measurements of air speed, ambient temperature at various heights from the ground, dry and wet bulb humidity, and mean radiant temperatures in the immediate vicinity. A calculation based on these raw scores yields the Effective Temperature, or New Effective Temperature, as it is now known (Goldman 1978).

Individual variability in thermal comfort is not easy to predict from these measures. In addition to ambient thermal conditions, a person's body warmth is a function of his or her clothing and level of activity. Methods of calculating the effects of these two variables have been incorporated into the Thermal Comfort Equation (Fanger 1972). However, more recent studies suggest that people's thermal comfort may be as much a function of the psychological power of suggestion, or of user information and control, as of measurable dimensions of the physical environment (Stramler et al. 1983; Wyon, 1988).

Thermal comfort curves have been derived from the results of laboratory studies where subjects have been required to rate or vote on various combinations of temperature and humidity conditions, ranging from most comfortable to least comfortable to actual interference with the task (MacFarlane 1978). These findings have allowed a comfort range to be prescribed for building interiors which specifies temperature and humidity combinations for various types of tasks. More recently, a calculation for air speed has been included in the formula to take into account the thermal effect of air movement in a space. High humidity means people experience temperatures as warmer than they do in drier air; rapid air movement means that people experience warm temperatures as cooler than they do in sluggish air. Instruments can and are used to measure these conditions in various work environments, and to determine, using charts and some simple calculations, how many of the people in a given environment can be expected to be "comfortable" thermally for how much of the time. The ASHRAE thermal comfort standard is based on these specifications.

ILLUMINATION

A thermal comfort-type approach to measuring visual or lighting ("luminous") comfort has been less satisfactory. Instruments are used to measure light output at the source, amount of light on the task and on the surround, amount of light absorbed by nearby dark colors or reflected by bright colors, and color rendition capacity of light quality, and various contract conditions. These measurements have not yet been systematically related to human peceptual judgments in a work situation, although much data collection and analysis has occurred (Boyce 1981). Calculations of the amount of glare caused by poor contrast conditions are complicated and are only partially related to the array of lighting influences on visual comfort (Hopkinson 1963). Most research on visual comfort has concentrated on variability in the speed and accuracy of human task performance according to the amount of light available. However, amount of light alone is coming to be considered a somewhat narrow definition of the range of influences on human visual comfort.

The amount of light in a given situation is a measurement easily taken with current instrument technology in artificially lit situations but is a more complicated calculation where daylighting is concerned. The amount of light emitted by various sky conditions is the preoccupation of daylighting researchers, and has been for some years (Flynn et al. 1979; Bryan 1983). Renewed interest in daylighting design has been generated by the realization that major energy savings can result from increasing the amount of natural light within a work space. The reactions of building occupants to window size and proximity, and their need for daylight for visual and health reasons, are critical issues in daylighting design (Vonier 1983b; Vischer 1987b).

The accessibility to measurement of *amount of light* from artificial and natural sources compared with other aspects of illumination quality has made it amenable to standard-setting. The "Equivalent Sphere Illumination" calculation, for example, is based on measured amount of light and has had widespread use as the standard of visual comfort in work settings. However, people tested in field settings do not always express increasing satisfaction with more light in the work place (Nemecek and Grandjean 1973; Boyce 1975). Studies have shown that eyestrain is a major source of occupant discomfort in office buildings (Stellman et al. 1982; Public Works Canada 1985). As many as 70 percent of respondents from one of the three office buildings described in chapter 4 reported some eyestrain at work. One thus infers that aspects of the visual environment other than *illuminance* (amount of light) must be affecting their visual comfort; for example, veiling reflections and other glare, "modeling" of three-dimensional visual fields, color rendition, and the *luminance ratio* between the visual task and the surrounding visual field.

Research is currently in progress toward the development

of a Visual Performance Index, which would represent a global score or calculation of the various effects of different lighting, daylighting, and color conditions on human behavior (much as the Thermal Comfort Equation represents the combined effects of temperature, humidity, and air speed on human thermal comfort). However, such a precise instrument measurement of human visual comfort is not yet available (Heerwagen and Heerwagen 1984).

ACOUSTICS

Acoustic conditions, although generally agreed to affect office workers, are less well-understood than illumination in office settings. Methods of measurement are cumbersome and detailed compared to the rapidity and sensitivity of the human ear and brain to sound. Typically, measurements of sound transmission through partitions, sound reverberation, and background sound levels are assessed against "noise criterion curves" to yield an assessment of acoustical comfort. Voice or speech privacy is calculated on an index betwen 0 and 1, where 0 is perfect privacy and 1 represents perfect communication conditions. The Articulation Index is based on how far and fast intelligible sound travels between two points in an unencumbered space.

Although noise (unwanted sound) is known to be a problem in offices, few acoustical standards pertain to white-collar work settings. Most acoustical standards exist to protect the human ear by indicating dangerous levels of noise in industrial environments. It is difficult to set comfort standards for a condition that may be either too loud *or* too soft for comfort! Noise that is too loud for comfort may be intrusive because it is a single and unexpected sound or because it is continuous and raises the overall level of noise in the work space (Warnock et al. 1972; Barge n.d.). Low background sound levels (noise that is too soft) mean poor acoustical masking of individual sounds like voices. As a result of low background sound levels, an office worker may experience stress from being too easily overheard or from being distracted by noise all day long.

Noise generated by human voices constitutes a different kind of stress from noise generated by equipment and machines, but the instruments available to measure acoustic performance in buildings do not correspond to the human experience of sound in this way. Sound transmission through walls and partitions, sound reverberation in a given space, and amount of sound (like amount of light) are all somewhat coarse measurements compared to the fine tuning of acoustic responses typical of people at work. For example, most noise measurements are unidirectional, that is to say, they measure sound between two points, assuming an emitter of the sound and a receiver, who hears it. For the person sitting at a desk at work, however, sound comes from all directions; the ear selects (voluntarily or not) sound from a number of different directions at once.

VENTILATION

Ventilation and indoor air-quality measurement are currently the least well-developed building performance disciplines in terms of predicting direct relationships between instrument measurements and human comfort. For example, considerable variation exists in standards of acceptability of various gases. Standards of *toxicity* have been developed for some airborne gases, but there are few standards of *discomfort* for most of the gases commonly found in indoor air (Levin 1984). For example, there is variation in acceptable levels of CO_2 in indoor air in offices, with comfort limits ranging from 400 to 1,000 ppm (parts per million), and dangerous limits ranging from 5,000 to 10,500 ppm. In Sweden a CO_2-driven ventilation system is available, whereby more air is delivered as CO_2 levels rise in the building (Sodergren and Punttila 1983). In North America it is more common for air handling systems to adjust the amount of air they deliver according to outdoor climate conditions, indoor room temperature, and operator response to complaints from occupants.

Methods of determining which gases (regardless of amount) are present in samples of indoor air are relatively undeveloped and awkward at this time. Instruments exist that pass samples of air through chemicals that will trap heavy molecules, and others that will trap light molecules. There is no way of knowing *exactly* which gas has been trapped; probabilities are estimated for the presence of various gases, depending on the profile of molecules found in the sample. Air-quality measurement in office settings includes, but is not limited to, analysis of various aspects of mechanical system performance, sampling and monitoring of CO_2 levels, testing for other contaminants such as respirable particulates, carbon monoxide (especially in buildings with parking garages and loading docks), ozone, and formaldehyde, and looking for biological contaminants such as mold, bacteria, or insect life (Turner and Bearg 1987).

Ventilation standards tend to regulate the percentage of fresh air introduced into the air distribution system at regular intervals (ASHRAE 1981). Some danger levels have been set by occupational health and safety organizations (e.g., NIOSH) for gases emitted in work environments, but almost no standards exist for ventilation comfort in nonindustrial work environments in terms of acceptable amounts of various pollutants in the ambient air. The human capacity for adaptation to smells and unpleasant odors makes it difficult to assess their subliminal effect on human comfort. The shortcomings of the ASHRAE approach result from the attempt to trade off energy efficiency with indoor air quality. Introducing less air into a building (and therefore not having to heat or cool it, humidify or dehumidify it) results in a lower energy consumption but causes a greater risk of air pollution or simply stale air inside the building. Criticism of the ASHRAE ventilation standard points out that the standard is based on 80 percent of occupants failing "to express dissatisfaction" with the air (Makower 1981). However, 1 or 2 or even 15 percent of occupants can be seriously affected by office contaminants (such as those emitted by photocopiers, printing equipment, new furnishings, and outdoor sources), and the building will still remain within the standard.

ERGONOMICS

Historically, the study of ergonomics was a study of man and his work. One important aspect of this was the work environment. The field of ergonomics includes *anthropometrics,* the study of the dimensions of the human body as related to machine and furniture design. Other aspects of ergonomics are the design of signs, instrument controls, visual displays, and lighting; *kinesthetics* (feedback to the body of muscle and limb movement), the design of man-machine systems, and the organization of work flow.

Originally applied to industrial environments, ergonomics is now a standard part of modern office design, especially with regard to the dimensions of furniture and equipment in the work space. Ergonomic issues have become critical where computers and other electronic equipment are used, especially for work groups who use VDTs. "Ergonomically designed" furniture is advertised for VDT users to help regulate their posture, arm and hand positions, eye-screen distance, neck and shoulder movements, and other critical dimensions of the electronic office.

Ergonomic standards are also available for desk height and depth of surface, chair back position, seat depth, screen height, keyboard placement, and other key components of an office work space where routine tasks are performed for long periods of time. Location of other visual material, the space, distance, and flow of products and tasks among related work groups, and the size of different work spaces are also aspects of ergonomic study in the office. The application of these standards results in more efficient work flow and use of human resources.

Ergonomic measurements rely on tape measures, individual task analysis, and systematic work-flow analysis, and can include measures of lighting and thermal comfort.

HOW DIAGNOSTIC BUILDING MEASUREMENT USES OCCUPANT RATINGS

Conventional approaches to building performance measurement of environmental conditions illustrate the value as well as the limitations of instrument measurements in determining comfort conditions in the work place. Three of the major limitations are:

- the difficulties of transposing laboratory-controlled conditions to field settings;
- the space and time limitations on instrument measurements in a building, as compared to human users' ability to judge environmental conditions within a broad spatial and temporal framework;
- the multiple and interactive effects of various environmental conditions on people's comfort judgments on their work place.

These are discussed below.

First, conventional building performance measurement is based in large part on laboratory research. Scientists systematically alter a given set of environmental conditions in the laboratory and require experimental subjects to perform typical tasks so that their speed and accuracy can be assessed under the various conditions. In other laboratory studies, subjects are typically asked to provide comfort and satisfaction ratings, such as "this is just uncomfortable," of various environmental conditions. When applied to field studies of human comfort and performance, the measurements yielded by the conventional approach are not always useful indicators of the experience of the worker in the office environment, where environmental conditions cannot be so rigorously controlled.

Second, under field conditions most instrument tests are likely to be spot checks, whether they are air samples, thermometer readings, or sound pressure levels. Some sample sites can be retested, but there is a physical limit to the amount of data that can be recorded per sample site. Occupant surveys, on the other hand, ask respondents to rate how comfortable or satisfied they have been over the whole period of time they have worked there, under a range of environmental conditions. As a result, each individual's response represents a *summary* of that person's experience of that condition (e.g., heat, light, odors) in that building over the time he or she has worked there. Occupants typically do not, unless specifically asked to do so, pause and assess the heat, light, and odors for *that particular moment in time* that coincides with the instrument spot checks.

It is therefore difficult to conclude that people's ratings of noise (for example) and the instrument testing of noise levels are measuring the same thing.

It is also difficult to control the effects of extraneous variables on instrument measurements in a building, and it is therefore even harder to know exactly what one is measuring. It is difficult to determine, on the basis of carefully planned, very thorough, and well-integrated spot-check data, whether the building, at the time it was tested, was typical for all other times, or just for some other times, or not typical at all. If the measurements were taken in summer, they may need to be taken again in mid-winter. Building operators report that the "changeover" seasons (spring and fall) are the hardest times to manage air handling, for example, and the times when they receive the most complaints from users.

This is not to say that instrument measurements cannot be integrated with occupants' ratings of environmental conditions but that this can be done only in specific and carefully defined ways. Building managers should remember that occupants are superior "measurers" to instruments, as they can provide an instantaneous summary of the *long-term performance* of the building environment, whereas instruments can only measure the here and now. Engineers tend to overlook the value of users' judgments because their assessments have not been validated by instruments. In fact, if instruments are useful at all, it is to help diagnose and clarify the a priori judgments that people have made.

The third limitation on building-performance diagnosis by conventional means is that instruments do not and cannot account for the *interaction* effects of environmental conditions on the perception or the task performance of the occupant. For example, a building may be experienced as hot by occupants because it is on a noisy street: people may not open their windows to reduce heat gain because of traffic noise (Canter 1975). It is well known by building managers that people in offices often experience the hiss of a background noise (sound-masking) system as evidence that air is being supplied by the ventilation system. Engineers know that people experience poor indoor air quality as "warm," and psychologists know that workers also confuse too much light with overheating (Nelson et al. 1984).

THE FALSE LOGIC OF INTEGRATING INSTRUMENT AND HUMAN ASSESSMENTS OF ENVIRONMENTAL CONDITIONS

One of the goals of a conventional building evaluation is to ally in a systematic way the instruments' measurements, their relationship to inferred comfort through standards and calculations, and the occupants' "subjective" feelings (or votes) of satisfaction. In evaluation studies where building occupants are asked for their assessments of the environment, it is usual to apply a conceptual framework in which instrument measurements are used to try to "harden" soft measurements such as rating scales. If an instrument shows that the temperature is 80°F in an office people judge as "too warm," then the people are "right" in complaining. If, however, an instrument shows that the ambient noise level is 32 decibels (DbA) in an office where people complain of noise—typically ambient noise levels indoors are 40-50 DbA—then the people are "wrong," and they are told to stop complaining.

The underlying assumption is that the "truth" in the "instrument comfort = human comfort" equation is the score yielded by the instrument, whether it is a thermometer, a sound pressure meter, or a photocell. Occupants' judgments and attitudes are dismissed as "subjective" and are typically disregarded by building designers, engineers, and psychologists in favor of more "objective" forms of measurement. Building studies aim to infer a logic to occupants' judgments that enables human judgment to be corresponded systematically to one or a series of instrument measurements. However, when it becomes apparent that there is no magical moment at which an instrument can be said to yield a precise measurement of human comfort, the *human* side of the equation—people's subjectivity, their unpredictability, and the problems inherent in measuring their behavior—are usually blamed for this failure. Typical complaints about people's judgments of environmental conditions include the following.

- *Unreliable.* Many psychological and social factors affect occupants' judgments of their work environment. Existing techniques of eliciting subjective responses do not *prove* that a consistent pattern of judging a condition as comfortable or uncomfortable exists. In other words, the logic of people's choices is concealed. Inconsistency in human judgments is an indicator of their unreliability (e.g., Rea 1982);
- *Subjective.* Time of day, time of year, type of task, and work organization affect the environmental condition being judged as well as the person's judgment. It is difficult to separate all these factors and identify the "true" environmental state of the building;
- *Unpredictable.* Only with difficulty do scaling techniques yield disaggregated occupant judgments. In "real world" building environments, workers are experiencing lighting, heating, ventilation, and acoustic conditions as a single event, and one does not know to which "stimulus" the persons' rating is a "response";
- *Unsystematic.* Finding out how much comfort building occupants feel does not provide an indicator of the point at which their *discomfort* might affect their productivity. To correspond to "hard" measurements, this should be a linear relationship, but most evidence suggests that it is not.

This list of concerns is valid if one accepts the assumption that underlies it: that people's attitudinal judgments of their environment require validation by instruments or standards to be considered "true." This assumption is prevalent in the scientific tradition. In chapter 3, "reductionist" studies were cited that specifically examined ways of measuring a person's perception of something as separate from ways of measuring the thing itself. This again raises the philosophical question of whether or not something perceived by people is as "true" as the same thing measured in another, nonhuman way. If there is no systematic way of fitting the two together, one measurement is taken as valid (i.e., the instrument test) *at the expense of* the other (i.e., the user survey).

Workers in office buildings have a hard time making their complaints heard by managers because everyone, including the workers themselves, is waiting for the complaint to be validated by some kind of measurement. The point of the building-in-use approach is to make something constructive and useful out of the complaints (or assessments) themselves. It may be philosophically appropriate to assume a separation between human perception and building performance, but in the real world of building management it is neither useful nor practical.

The purpose of the building-in-use approach is to demonstrate that human judgment alone can provide an adequate and useful measure of building environments. Its intrinsic subjectivity and "softness" is, in fact, its greatest asset, and one that should be capitalized upon by building researchers, owners, and managers. The assumption behind the building-in-use approach is that *intrinsic qualities of the environment are in fact being measured by occupants' ratings.* Rather than try to relate people's judgments of warmth in a building to a thermometer reading, one can take the warmth rating and the thermometer reading as two separate, if related, *but equally valid* measures of that environmental condition. This also dispenses with the extra step of making the instrument reading "meaningful" by calculating its relationship to derived comfort parameters.

HOW THE BUILDING-IN-USE APPROACH USES OCCUPANT RATINGS

In employing the building-in-use approach to environmental evaluation, one needs a valid and reliable set of rating scales for occupants' subjective assessments of the environment, but one does not need a large range of complex and expensive instrumentation as one does for conventional building performance evaluation. Occupants' ratings can be integrated to provide a profile of building-in-use performance based on the users' experience of the building. The "truth" about the building lies in this elicited experience of the users and not exclusively in what the instruments tell us.

An example of the building-in-use approach is the following interpretation of data from the Central Building. As noted in chapter 4, this building's occupants were surveyed twice at different times of the year. On the second occasion the building population had nearly doubled because of the introduction of a large number of temporary clerical workers. It was expected that occupants' environmental ratings would be worse on the second occasion because the building was relatively crowded at this time. Using conventional building diagnosis, one would predict that the increased number of workers resulted in less comfortable conditions—such as more noise, less storage space, and poor ventilation—and thereby more occupant discomfort. If the crowded conditions were affecting building performance adversely, the second user group should have judged ventilation more poorly than the first. But in fact the second occupant survey yielded a *better* rating of ventilation conditions than the first. This means either that the building systems perform better than one might conceivably have anticipated, or that people are so changeable that there is no reliability in the ratings.

In adopting the latter argument one might emphasize that the second group of users are temporary, short-term and well paid. They are therefore predisposed to be less critical of the building's performance than the permanent employees, who are there year-round. Using the diagnostic approach, one might take measurements of ventilation at the time of the first user survey and compare them with ventilation as measured at the time of the second, thereby determining with finality whether this difference in user assessments was a function of mechanical system performance or of human unpredictability.

The problem here is to decide *how* to measure ventilation:

Air circulation tests? Air sample analysis? Particulates measurement? Effective ventilation rate? Percentage of fresh air in the air supply system? Carbon dioxide measurements? Or some combination of the above? Moreover, how frequently should such measurements be taken? Once? One hundred times? Every day for a week? Morning and afternoon? One year and five years after occupancy? Each one of the decisions made about how, where, and when to take instrument measurements of ventilation quality constitutes a *narrowing of the definition of the meaning* of ventilation. This process causes the results of the measurements to shift away from the broader and more wide-ranging experience of the occupants judging their experience of ventilation conditions in the building. Moreover, the ventilation measurements will be inconclusive because of the difficulty of isolating the conditions under study in the field. Even if evidence of an objective difference in ventilation emerged conclusively from the data, there would still be questions of changing climatic conditions and variation in systems-operating practices that contaminate the quasi-experimental design and prevent clear conclusions being drawn regarding "actual" or "true" ventilation quality differences between the two conditions.

The building-in-use model, unlike the conventional diagnostic model, uses the same information to conclude that the building's ventilation is, in fact, better on the second than on the first occasion *for no reason other than that the occupants judge it so.* Whether they judge it so because they are temporary workers or because a larger amount of air is in fact supplied by the systems under crowded conditions is only of speculative interest; it does not affect the usefulness of this finding. Should the building's ventilation be rated very negatively by occupants, one might equally well, in order to effect an environmental improvement, increase the proportion of well-paid temporary workers as retrofit the mechanical system. Assuming that the rating scales used to elicit occupants' responses are reliable, the prediction we can make is that office building ventilation systems perform better for well-paid temporary occupants than for other groups of clerical workers. This could have profound implications for the selection of employees, organization of tasks, and matching of work groups to buildings.

The building-in-use paradigm incorporates gender differences in building occupant judgments in the same way. The Eastern Building described in chapter 4 is a good example of how a floor dominated by female workers performs less well in spite of extensive physical improvement than a conventional floor with even numbers of men and women. One can conclude that buildings that expect to be occupied by a majority of female employees will have to be designed to a higher standard to create in users an equivalent satisfaction level to that of men working in an average state-of-the-art office environment.

BUILDING-IN-USE MEASUREMENT OF ENVIRONMENTAL QUALITY

It is apparent from this discussion that the ratings elicited from occupants are crucial to the building-in-use assessment. Not just any set of scales will do. It is important to know how many to use, how they should be worded, what they should ask, and whether there is a finite number of ratings for each environmental condition.

Many building evaluations have used occupant surveys to get at users' assessments of the environment, and a preponderance of these has used some kind of rating scale technique. There is no apparent reason why any one of these scaling techniques is better or worse than the others. However, regardless of the *type* of scale used, it is important to remember that the various diagnostic categories of building performance comprise several rating scales: one rating alone is not sufficient to enable a user to judge his ambient environment. It is even more important to ensure that the same set of scales is applied to every building assessed. If occupants of different buildings (or different floors of a building) are supplying different ratings, then there is no basis for comparison, and no database can be used for the development of norms.

Whatever type of rating scale is used, one must be sure that one is measuring those environmental conditions that represent the salient dimensions of occupants' experience of their building. It is pointless, for example, to get a rating of whether or not background sound levels are low enough to be unintrusive if the occupants' concern is that they be high enough to mask conversation. And it is not useful to ask people whether they are comfortable with the amount of humidity in the air if they are constantly complaining to each other about the dryness. The interactive effects of lighting, noise levels, and temperature in offices should also be considered in defining useful rating scales. Environmental conditions other than the diagnostic categories of indoor air quality, lighting, thermal comfort, and acoustics may also exist that are important ingredients of users' experiences of the office environment.

In the next chapter the building-in-use criteria of environmental quality are discussed. We point out which ratings are the crucial ones, corresponding most closely to the actual way in which people experience buildings—in other words, the critical psychological dimensions. Just as being able to open windows is more psychologically important to people in office buildings than "behind the scenes" changes to the mechanical systems, there are other psychological ways in which people assess their work environment. We will compare these psychological ways of experiencing the environment to the diagnostic categories of instrument measurement that were explored in this chapter.

SUMMARY

The building-in-use profile provides scores on the critical dimensions of building-in-use performance. These scores are computed from occupants' ratings of a number of environmental conditions on a scale of 1 through 5. The scores are numerical indicators of environmental quality. They are informative in themselves and can also be compared to the normative scores for a large number of buildings. This instantly useful diagnostic approach to the question of "Is this a good building?" is an improvement on the existing pattern of diagnostic assessment in conventional building-performance categories.

In traditional building-performance diagnosis, instruments have been used to measure temperature, indoor air quality, acoustics, illumination, and the ergonomic environment to determine whether or not these met preestablished standards of safety and comfort. In the building-in-use approach, instrument measurements can play an important follow-up role in adding to occupant information, but this should not be confused with validation of occupants' ratings. The diagnostic categories of building performance represent dimensions of building quality that are imposed from the outside in the form of standards of comfort, health, and safety, and are limited by the range and type of instrumentation available to take measurements. Building-in-use assessment systematically measures the quality of the environment as it is experienced by building users. This assessment yields an understanding of quality according to a system of logical inference that is intrinsic to the performance dimensions themselves and *not* imposed from the outside.

By offering a quantifiable answer to the question of whether or not an organization owns, occupies, or is operating a "good building," the building-in-use assessment may help building managers make a better case to senior management for resources for building improvement. Rather than being forced to react to occupants' complaints ("e.g., We need $10,000 to relocate the thermostats in this building, because people cannot regulate their heat and cold . . .") and having to prove that features of the building environment affect worker productivity, managers can compare environments to each other on a numerical scale of "goodness" and argue that a self-respecting corporation does not want to operate a less-than-good building. There is no absolute truth about the best building; there is only degree of environmental quality compared to other, comparable buildings.

The rating scales used in a building-in-use assessment are critical. They do not necessarily correspond to conventional categories of building performance diagnosis. In the next chapter we learn something about the ways in which people judge the interior work environment, and how these psychological dimensions of the building-in-use are an intrinsic part of the process of assessing environmental quality.

Chapter 6

THE BUILDING-IN-USE DIMENSIONS OF ENVIRONMENTAL QUALITY*

Because the assessment of environmental quality focuses on the users' experience of their building environment, it is necessary to determine how people themselves actually assess their ambient environment—their surroundings. While some psychological experiences correspond to the diagnostic categories of comfort measurement described in chapter 5, others, as we shall see, do not. The disparity between conventional definitions of ventilation, thermal comfort, and acoustics, and users' environmental experience of air quality, temperature, and noise, explains in part why an office that meets ASHRAE's environmental standards does not always guarantee good environmental quality.

In this chapter we will explore the various ways in which people actually experience their environment and compare and

* The statistical procedures used to elicit these dimensions and the interpretation of statistical results were carried out under the direction of Dr. Richard Dillon of Carleton University, Ottawa.

contrast these to conventional environmental measurement. The discussion falls into four parts. The first defines seven building-in-use dimensions, or ways in which people experience their ambient surroundings; this part describes how the dimensions of environmental experience were arrived at and what they mean. The second, third, and fourth parts discuss how these seven dimensions of building-in-use performance can be used to predict worker morale, productivity, and physical well-being.

THE SEVEN BUILDING-IN-USE DIMENSIONS OF OFFICE ENVIRONMENTAL QUALITY*

The seven dimensions uncovered by the data analysis represent seven categories of users' environmental judgments. They are:

- Air Quality
- Noise Control
- Thermal Comfort
- Privacy
- Lighting Comfort
- Spatial Comfort
- Building Noise Control

Each category is measured by the same scales as those used to assess the conventional ambient environmental conditions of thermal comfort, ventilation, lighting, acoustics, and ergonomics in the office buildings that were studied. But the building-in-use dimensions do not correspond to conventional building performance categories: the ratings group according to a different logic and in a different way for people's experiences than they do for diagnostic measurement.

Ratings from close to 3,000 workers in five government office buildings were elicited on scales that were designed to

---

*See Dillon and Vischer (1988) for a complete technical report of the statistical procedures for analyses of these data. They are summarized in the Appendix of this book. They are drawn from an in-depth statistical analysis of survey data from the buildings described in chapter 3 and some others.

correspond to the ambient environmental conditions that are typically measured by diagnostic instruments to assess human comfort in the work environment. However, when they were analyzed, the ratings did not cluster as expected. Instead, they emerged regrouped in such a way as to suggest dimensions of occupants' experience of ambient environmental conditions in the work place that do not correspond to the conventional categories.

These seven building-in-use dimensions are the generic criteria for office environmental quality. These alternative, user-based dimensions of building performance incorporate interactive effects, individual psychological differences, and the varying effects of individual spatial location in the building. They are the building-in-use criteria for assessing environmental quality in offices. The building-in-use assessment system for evaluating office interiors uses the norms from these seven dimensions to generate a *building-in-use profile* for all or a part of an office building. The scores on the seven dimensions more closely represent the quality of occupants' experience than any other type of building performance measurements; these dimensions are the criteria for assessing the users' experience of the office environment. For the rest of this book, they will be referred to as the *building-in-use criteria for, or dimensions of, office environmental quality.*

The seven criteria are described below, along with the scale ratings that comprise each one. They are ordered from most to least clear-cut and identifiable dimension of environmental quality.* For each dimension the ratings that comprise it are presented in their order of importance, and some time is spent discussing the meaning of each regrouped and new way of clustering these ratings. In discussing the meaning of these regrouped clusters of ratings, we indicate which are the most important components of the dimension, that is, those from which an overall score for each dimension can be predicted. Each is given a building-in-use name, but this should not be confused with the conventional meaning of each term. This name has been selected to denote the meaning of each

---

*Each criterion emerged as a factor from a Factor Analysis of thirty-five environmental rating scales (See Appendix).

Air Movement:	1	2	3	4	5	(0.88)*
	"Stuffy"			"Circulating"		
Air Freshness:	1	2	3	4	5	(0.85)
	"Stale Air"			"Fresh Air"		
Ventilation:	1	2	3	4	5	(0.79)
	"Bad"			"Good"		
Odors:	1	2	3	4	5	(0.50)
	"Unpleasant"			"Not Noticeable"		
Humidity:	1	2	3	4	5	(0.47)
	"Too Dry"			"Comfortable"		
Warmth:	1	2	3	4	5	(0.40)
	"Too Warm"			"Comfortable"		

* Varimax rotated factor loading, from a Maximum Likelihood Factor Analysis.

6-1. Air Quality ratings, in order of importance.

dimension clearly, regardless of any other meanings the words themselves might have.

AIR QUALITY

Air Quality is composed of six ratings, listed in order of importance in figure 6-1. Air Quality is a particularly interesting dimension because ratings show that users emphasize air freshness and circulation over simple judgments of whether the ventilation is good or bad. This combination of ratings is consistent across all the buildings, indicating that this is a stable and reliable category of environmental quality.

In instrument-based building diagnostic systems, the degree of warmth that is comfortable to users is considered a thermal-comfort judgment rather than an air-quality judgment and so is the degree of humidity (or dryness) of the air. However, in the building-in-use system, the experience of warmth and humidity of the air is part of people's sense of Air Quality, as is whether or not people are aware of unpleasant odors in their work space.

The building-in-use dimension of Air Quality emerged strongly from the analysis and is the most pronounced dimen-

	1	2	3	4	5	
General Office Noise Levels:	1	2	3	4	5	(0.87)*
	"Too Noisy"			"Comfortable"		
Specific Noises of Voices and Equipment:	1	2	3	4	5	(0.82)
	"Disturbing"			"Not a Problem"		
Noise Distractions:	1	2	3	4	5	(0.75)
	"Bad"			"Good"		

* Varimax rotated factor loading, from a Maximum Likelihood Factor Analysis.

6-2. Noise Control ratings, in order of importance.

sion of users' experience of their environment. The entire constellation of ratings that constitutes building-in-use Air Quality can be predicted from the first three ratings. This is useful to know for the study of future buildings.

NOISE CONTROL

Noise Control, the second dimension of occupants' environmental experience, is interesting because it does not incorporate conventional dimensions of the acoustic environment, such as acoustic privacy and communication conditions. In diagnoses of building performance, these are routinely measured as part of the acoustic package. The analysis shows that people in offices judge their acoustic environment more on the basis of how much they are aware of and disrupted by noise. The components of this dimension, listed in order of importance, are shown in figure 6-2.

Interestingly, the dimension of Noise Control is separate and distinct from occupants' acoustic perception of noise created by features of the building—Building Noise Control, a separate building-in-use dimension described below. This is useful to know in practice, because different actions are required to solve different types of noise problems. A poor rating of the noise that people make—noise intrusion—requires a different sort of intervention than a complaint about the noise

Amount of Space in Your Work space:	1	2	3	4	5	(0.65)*
	"Bad"				"Good"	
Work Storage:	1	2	3	4	5	(0.65)
	"Adequate"				"Insufficient"	
Furniture Arrangement in Your Work space:	1	2	3	4	5	(0.59)
	"Bad"				"Good"	
Personal Storage:	1	2	3	4	5	(0.56)
	"Adequate"				"Insufficient"	
Furniture Comfort:	1	2	3	4	5	(0.44)
	"Bad"				"Good"	

* Varimax rotated factor loading, from a Maximum Likelihood Factor Analysis.

6-3. Spatial Comfort ratings, in order of importance.

levels of buzzing lights and other such building-generated noise sources—building noise.

Although many other acoustic ratings were tested, these three are the most important predictors of the Noise Control rating.

SPATIAL COMFORT

Spatial Comfort represents that area of human comfort conventionally called *ergonomic comfort* and is related to furniture design and its placement in the work space. This dimension includes the scales listed in figure 6-3.

The most interesting feature of occupants' experience of Spatial Comfort in the work space is that it emerges so clearly as a cluster of related judgments. Although the importance of furniture and layout is not underestimated in most offices, little is known systematically about how ergonomic factors affect the satisfaction and productivity of office workers.

Design decisions such as size of office or individual work station, arrangement of walls or screens, how many (or how few) work stations should be enclosed in one space, and whether or not people can see their coworkers and be seen by

Cold Temperatures:	1	2	3	4	5	(0.77)*
	"Too Cold"			"Comfortable"		
Temperature Shifts:	1	2	3	4	5	(0.61)
	"Too Frequent"			"Generally Constant"		
Temperature:	1	2	3	4	5	(0.60)
	"Uncomfortable"			"Comfortable"		
Drafts:	1	2	3	4	5	(0.58)
	"Drafty"			"No Drafts"		

* Varimax rotated factor loading, from a Maximum Likelihood Factor Analysis.

6-4. Thermal Comfort ratings, in order of importance.

them, are all aspects of this dimension. Designers and space planners know that furniture orientation and layout of the work space affect users' experience of thermal, visual, and acoustic conditions. It seems that users themselves may be more aware than they are usually given credit for of the importance of the physical layout as a mediator of their environmental experience.

The first four of these ratings are those that best predict the building-in-use Spatial Comfort score.

THERMAL COMFORT

Although the name assigned to this building-in-use dimension is the same as the name of the traditional thermal comfort discipline, of the ASHRAE standard, and of conventional tests of building performance, the components of building-in-use *Thermal Comfort* are clearly different from the above. Its major component is coldness rather than warmth, as figure 6-4 shows. Occupants of office buildings assess Thermal Comfort in terms of whether they feel cold, whether they feel drafts (intermittent coldness), and whether they are bothered by temperature fluctuation. Conventional thermal comfort models measure feelings of warmth and humidity; the building-in-use model treats these two items as aspects of Air Quality.

This particular selection of ratings indicates that where room temperature is concerned, occupants are aware of cold and drafts as sources of discomfort. Because overly warm air is also likely to be stuffy and dry, warmth is included in their experience of ventilation, or Air Quality. Occupants relate

	1	2	3	4	5	
Electric lighting:	1	2	3	4	5	(0.67)*
	"Bad"			"Good"		
Glare from Lights:	1	2	3	4	5	(0.66)
	"High Glare"			"No Glare"		
Brightness of the Electric Light:	1	2	3	4	5	(0.55)
	"Too Much Light"			"Does Not Get Too Bright"		
Colors:	1	2	3	4	5	(0.47)
	"Unpleasant"			"Pleasant"		
Daylight:	1	2	3	4	5	(0.38)
	"Bad"			"Good"		

* Varimax rotated factor loading, from a Maximum Likelihood Factor Analysis.

6-5. Lighting ratings, in order of importance.

Thermal Comfort to temperature constancy or the lack of it, and probably, therefore, to being able to adjust temperatures to suit their own individual thermal conditions, which change according to such random factors as time of day, food intake, and degree of emotional arousal. The building-in-use Thermal Comfort score can be predicted from the first three ratings.

LIGHTING COMFORT

Lighting Comfort is different from the other building-in-use dimensions inasmuch as the conventional diagnostic criterion of lighting measurement (amount of light) and the building-in-use criterion appear to coincide. Office occupants' building-in-use judgments indicate a sensitivity to too much light or brightness in their work space. They are also aware of color rendition. The major component scales of Lighting Comfort are ranked in importance as shown in figure 6-5.

Lighting Comfort does not emerge as clearly from the data analysis as do the previous four dimensions. One reason for this is that occupants may not be the best judges of their own lighting conditions. People's judgments of amount of light may reflect a judgment of quality or appropriateness of the lighting rather than their assessment of how much light they have. Moreover, in many modern office buildings, workers' visual comfort may be more compromised by overlighting than by insufficient lighting.

Studies of daylighting of office buildings have shown that users tend to overestimate the amount of natural light that reaches their work surface (Wells 1965; Manning 1968; Bryan, et al. 1988). The fact that a rating of daylight quality is part of users' lighting experience implies that window proximity and view as well as daylight as a light source are combined in occupants' assessments. This finding has implications for work-station layout and desk orientation as well as for window placement and design.*

Although amount of light or illuminance alone is not enough to guarantee lighting quality, it is still a widely used standard of assessing the visual environment. Most office lighting systems are designed according to standards that regulate the amount of light emitted by fixtures. The components of the building-in-use Lighting Comfort dimension, however, illustrate that lighting quality is assessed according to other visual factors such as color rendition, glare, and light from windows. The first three ratings will adequately predict the building-in-use Lighting Comfort score.

PRIVACY

Privacy, as building-in-use occupants experience it, incorporates acoustic and visual aspects of the physical environment. The components of Privacy, listed in order of importance, are shown in figure 6-6.

Occupants' experience of Privacy is interesting because it incorporates both acoustic conditions and work-station layout. Conventional acoustic analysis integrates noise levels with voice privacy (for example, by introducing background noise to protect voice privacy). The building-in-use assessment clearly does not. In fact, surveys of office occupants show

*It is interesting to note that in preliminary data analysis, users' judgments of the quality of daylighting in their work space emerged as a separate and distinct dimension. In it, the daylight rating was coupled with ratings of window proximity and glare from the windows. This should be borne in mind in understanding building-in-use Lighting Comfort scores.

	1	2	3	4	5	
Voice Privacy:	1 "Bad"	2	3	4	5 "Good"	(0.80)*
Telephone Privacy:	1 "Bad"	2	3	4	5 "Good"	(0.73)
Visual Privacy:	1 "Bad"	2	3	4	5 "Good"	(0.50)

* Varimax rotated factor loading, from a Maximum Likelihood Factor Analysis.

6-6. Privacy ratings, in order of importance.

that background noise is often perceived by occupants as intrusive and disturbing rather than as protective of voice privacy (see chapter 5)—a finding that makes sense only if one separates the experience of noise from the experience of privacy.

Visual privacy is also a component of the overall experience of privacy in offices. Occupants use elements of furniture arrangement and work-group layout to protect themselves from unwanted visual contact for certain aspects of their work, and to control their physical and visual accessibility to coworkers. Acoustic screens, which are usually employed in open offices to protect workers' voice privacy, are often drawn around three or even four sides of a person's desk in an attempt to protect visual privacy. In the building-in-use assessment, acoustic and visual privacy merge in the office worker's experience of the office environment. The three ratings listed here are sufficient to predict the building-in-use Privacy score.

BUILDING NOISE CONTROL

The building-in-use *Building Noise Control* dimension differs from the elements of Noise Intrusion; the ratings are listed in figure 6-7.

The existence of this dimension shows that occupants assess noise from building-related sources differently from the ways they assess noise generated by their coworkers' activities. Building noise constitutes a separate acoustic experience, and the ways in which it is disturbing differ from the ways in which "people noise" becomes intrusive. Occupants' tolerance for building noise may be different from their tolerance for people

	1	2	3	4	5	
Noise from the Lights	1	2	3	4	5	(0.68)*
	"Buzz/Noisy"			"Not Noticeable"		
Noise from the Air Systems	1	2	3	4	5	(0.55)
	"Disturbing"			"Not Noticeable"		
Noise from Outside the Building	1	2	3	4	5	(0.55)
	"Disturbing"			"Not Noticeable"		

* Varimax rotated factor loading, from a Maximum Likelihood Factor Analysis.

6-7. Building Noise ratings, in order of importance.

noise (or noise intrusion), regardless of the actual sound level of the noise sources.

When facilities managers hear office occupants' complaints about noise, they can only act on them effectively if they know which are building-related and which are not. The actions they will take to intervene will vary according to whether they are solving building noise or people noise problems. Although there are many additional sources of building noise in an office, the three ratings listed here will adequately predict the building-in-use Building Noise score.

All seven building-in-use dimensions of environmental quality emerged in all the buildings, which shows them to be stable and reliable categories or ways in which users judge office environments. They explain and expand upon the human side of conventional diagnostic building-performance measurement categories. They represent user-based criteria for building-in-use assessment. In examining, analyzing, understanding, and acting on these seven dimensions of environmental quality, one is truly formulating the prototype building-in-use for future offices.

SCORING THE BUILDING-IN-USE

The scores that can be computed for each building-in-use dimension represent a way of quantifying environmental quality. Because the building-in-use dimensions can be measured, environmental quality in offices can itself be measured. If the building-in-use dimensions are the core of the building-in-use concept, the computation of scores on the dimensions is the core of the assessment system. It is this process of computation that is discussed below.

As explained in chapter 5, it is possible to compute a score for each dimension, both in each building separately and for a number of buildings, by adding the ratings and calculating their mean or average. Each scale receives a rating between 1 and 5, inclusive. Each set of ratings is averaged to provide a score between 1 and 5 for each dimension. This number, which can be carried to one or more decimal places, is the indicator of quality on each building-in-use dimension.

When all the scores on a dimension are added and averaged across all buildings, we have a norm on each dimension for all buildings to which each individual building can be compared. These multibuilding norms are stored in a database. The scores on a single building can be compared to the norms to determine which dimensions are normal, which are below normal, and which are above normal. Figure 6-8 gives the norm or baseline score for each building-in-use dimension. These scores represent building-in-use norms for large, contemporary, multistory office buildings in northern climates. They are listed from "best" (closest to 5) to "worst" (closest to 1).

These norms are derived from "normal" (i.e., unproblematic) office buildings. The fact that none of the dimensions scores less than 2 suggests that most unexceptional buildings-

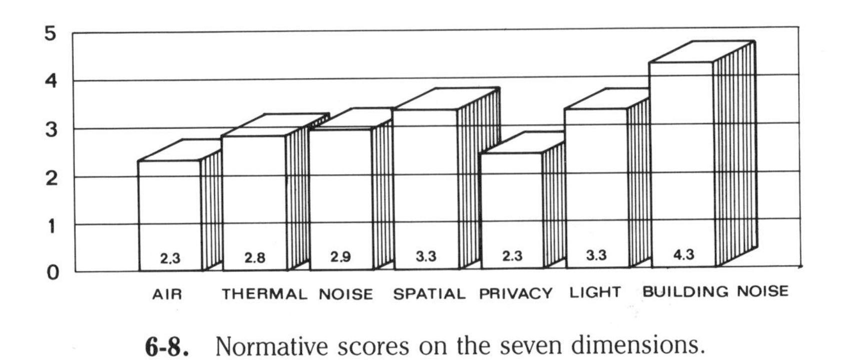

6-8. Normative scores on the seven dimensions.

in-use, like the ones reported here, perform at a level that most of the occupants find tolerable, if not fully acceptable. The fact that it is "normal" for Air Quality and Privacy to be as low as they are is an indictment of conventional office environments. These two norms confirm what is generally known about offices: that workers in sealed modern office buildings complain about poor indoor air quality, and workers in open-plan office spaces complain about lack of privacy.

Figure 6-8 shows that the highest-scoring dimension is Building Noise Control, with a score of 4.3. Ten points separate this rating and the next two, which are Spatial Comfort and Lighting Comfort at 3.3. There is a middle set of ratings that cluster just below 3, or average: Noise Control and Thermal Comfort. Thus, on the scale of 1 to 5, an office that scores 3 on Thermal Comfort, above 3 on Lighting Comfort, and below 3 on Air Quality is not just performing in an average sort of way for its users; it is also typical of North American office buildings. Just how "normal" is this range and ranking of building-in-use scores?

BUILDING DIFFERENCES

In figure 6-8 we see the norms for each dimension, which represent all buildings added together. The group of buildings from which this database was developed are all multistory office buildings, all fairly recently constructed, all owned and operated by the government, and located in different parts of Canada. These norms would not be representative of buildings used for manufacturing purposes, or for schools, or for buildings in Florida or Japan, for example (the former because of its warm climate, the latter because of its cultural differences). They are at least partially representative of privately owned office buildings, of small, old, or single-story office buildings, and of offices all over North America.

Differences among the individual buildings that comprise the database help us to understand something more about the buildings themselves and to determine which of the dimensions are most constant and stable, and which show variation.

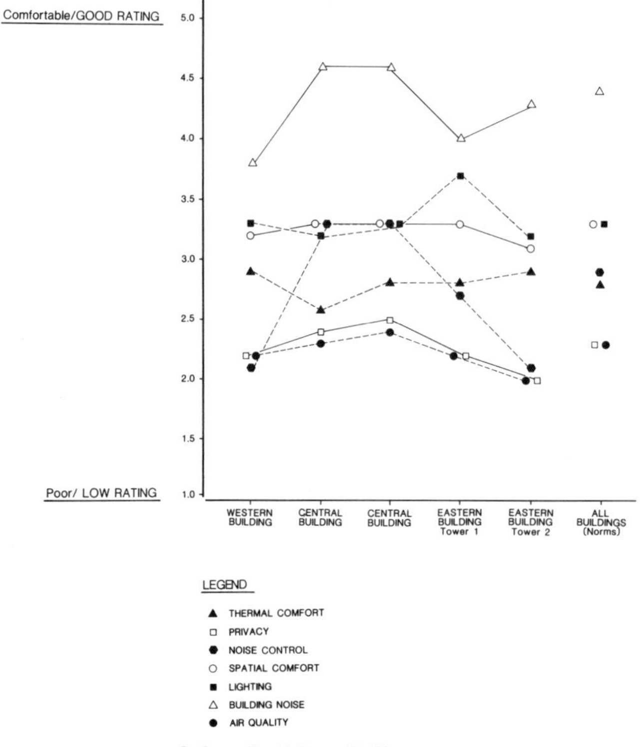

6-9. Building Differences.

The graph depicted in figure 6-9 shows differences among the buildings on the seven dimensions. By separating out building scores, it is possible to determine the items on which the buildings vary from one another. It is also possible to see the dimensions on which individual buildings differ significantly from the norm. The fact that there is so little variation among buildings supports the reliability of the measurements and the validity of the dimensions. The small variation that does exist shows that the seven building-in-use dimensions are consistent enough for the multibuilding scores to serve legitimately as norms for building-in-use dimensions.

The graph shows inconsistency across buildings on two of the seven building-in-use dimensions. These two—Noise Control and Building Noise Control—show some variation by building, a finding that suggests there was something specific affecting these building-in-use scores in the deviant buildings.

Figure 6-9 shows clearly that the Central Building has significantly higher ratings on both building-in-use noise dimensions than the other buildings. This means that this building is not judged as uncomfortable as the other two buildings with regard to various types of noise. The diagnostic instrument tests of noise in the Central Building indicated that noise levels in this building were not significantly different from those in the other buildings. What, then, causes the Central Building to score so well on the noise control dimensions?

What differed in the Central Building was the nature of the work people perform there. Their assembly-line paper-processing tasks make workers less sensitive to the overall noise level—regardless of the instrument-measured level of noise. Noise in that building was not judged to be intrusive to the performance of work in the way that it was in the Western and Eastern buildings. Central Building occupants indicated by this judgment that they do not need low noise levels to concentrate; they do not make confidential telephone calls or have confidential conversations as part of their job. They sit in large open areas without enclosure or with group work-space enclosure, and they work together as a group. They do not suffer from noise control problems. The level of noise is perceived as appropriate for the type of task being performed. In the other buildings, where more conventional office tasks require work-station enclosure, where the telephone is used, where meetings and conversations are part of the job, and where intrusive noise levels are a real interruption, building-in-use noise control ratings, especially for people-generated noise, were significantly lower.

Building-in-use exceptions to the norm are important because the reason for establishing norms is to be able to earmark exceptions and explore them further in the context of a specific building. In addition to the noise control ratings, figure 6-9 indicates that building-in-use Lighting Comfort, though rated well, is not rated equally well in all buildings. Occupants in the Western Building rate the Lighting Comfort dimension better than Central Building occupants. Why? Investigation showed that the layout of offices in the Western Building responds to the large number of windows and relatively shallow floors. In general, occupants of the Western Building have good daylight. Central Building occupants also have good daylight, because of the absence of partitions and walls. However, their rating shows they do not really value it, probably because of the high rate of VDT use in that building. Moreover, although the electric lighting system is essentially the same in both buildings, the visual task requirements of Central Building occupants are far more demanding than those of Western Building occupants. The reading and sorting of tax returns is illuminated by a uniform, high-level overhead lighting system that is more suitable to the conventional office tasks carried out in the Western and Eastern buildings.

So whereas Central Building occupants are *less* concerned about noise because of their task requirements, those same requirements cause them to be *more* sensitive to lighting conditions. However, this should not cause the building-in-use scores to be viewed as judgments of task alone. In assigning these ratings to environmental conditions, people are judging the environment in which the task is being carried out. In fact, it is impossible to separate how much of the occupants' judgments are related to the task and how much to the environment in which they carry it out. It is the interaction and the degree of mutual suitability between the building environment and the jobs people are doing in it that is the strength and unique contribution of the building-in-use assessment approach. A single measurement—the building-in-use score—is an assessment of this interaction; in this way, building-in-use assessment is a methodology that integrates rather than separates, that measures the whole rather than the parts.

THE RELATIVE IMPORTANCE OF THE BUILDING-IN-USE DIMENSIONS

Are the seven critical building-in-use dimensions all equally important to a person working in an office, or are some more important than others? Which of these dimensions, in other words, are the most critical to worker productivity and morale? A manager needs to know whether Spatial Comfort has the same degree of effect on worker performance as Air Quality, or whether she should be worrying about Thermal Comfort more than Noise Control to get people to work well in their offices.

The previous discussion noted that Building Noise Control was a highly rated dimension in office buildings; that is, the norm for Building Noise Control is over 4 and higher than all the other scores. In this section, we will answer both the question of whether or not this high norm means that managers should rush to control building noise in their buildings; and the question of whether the fact that it scores highly means it is more important to worker performance than the other dimensions. For example, Building Noise Control may contribute to a good-quality environment, but a decision maker needs to know how important this feature is to occupants as they carry out their daily tasks. Do privacy and window access, for example, affect people's work behavior more directly than buzzing lights or noise from the air system?

To answer questions like these, we carried out a further stage of data analysis to determine which of the building-in-use dimensions are most important to the way people do their work in offices, to their comfort and satisfaction, and to their productivity.* Some building-in-use dimensions affect how much occupants *like* the space; others affect how well they can *work* in it. Satisfaction with an environment—and being able to function effectively within it—are related but not identical.

To determine how effectively the seven dimensions of environmental quality predict workers' *satisfaction,* a statistical model was constructed, using responses to the question, "Compared to other work stations in this building, how do you rate your individual work station on a scale of 1 (poor) to 5 (good)?" Workers' environmental satisfaction is directly proportional to their psychological *morale.*

The analysis also examined the relationship between building-in-use scores on the seven dimensions and worker *productivity,* as measured by the responses to the question, "Please estimate how well or badly this space affects your ability to do your work on a scale of 1 to 5, where 1 is "Makes it

* A Least Squares Regression analysis was used. The statistical analysis techniques are summarized in the Appendix.

TABLE 6-1
ENVIRONMENTAL SATISFACTION AND WORKABILITY IN OFFICE BUILDINGS

Building	Satisfaction (Morale)	Workability (Productivity)
Western Building	3.1	3.1
Central Building	3.1	3.4
Eastern Building Tower 1	3.2	3.0
Eastern Building Tower 2	3.3	3.4

difficult" and 5 is "Makes it easy." The rating that occupants supplied in response to this question was considered a score of how "workable" the work space is for the occupant. The *workability* of the environment, it turns out, is directly proportional to the *performance* of the worker. Workability is correlated with environmental satisfaction, but it is not the same thing.

Statistical analysis of occupant responses shows that people's work-station satisfaction scores (their level of psychological morale) and office workability scores (their individual productivity) in all the buildings studied is around 3, or average. The scores are presented in table 6-1. As this table shows, variation among these office buildings is minimal, suggesting that most office occupants generally feel that they can function satisfactorily in their work space. Generally, office space in these buildings is considered slightly more workable than satisfying.

USING BUILDING-IN-USE SCORES TO PREDICT MORALE

Worker satisfaction can also be labeled *morale* because the degree to which someone does or does not like the space he works in has a definite effect on his state of mind and general outlook on his work. Working in a very nice office can compensate for dissatisfactions on the job, in the same way that a very satisfying job can be vulnerable to the small but persistent annoyances of an unpleasant work space. Managers recognize this when they use pleasant office space as a reward for promotion or advancement in an organization.

How well do the seven dimensions of the building-in-use system predict worker morale, and how much do they affect morale relative to other, random factors in the situation, such as individual differences, time of day, and organizational behavior? The statistically derived answer is that the building-in-use dimensions account for some 25 percent of the variability in workers' satisfaction. This is quite a high proportion, leaving 75 percent to be accounted for by the kinds of social and psychological differences listed above as random factors. Conventional wisdom does not ascribe this much influence to the physical environment: it is usually discounted in managers' assessments of what makes people feel (or not feel) better at work. If 25 percent is applied to a dollar calculation of what is spent on people's salaries, it is a sizable investment. The 25 percent increase in environmental satisfaction that might be brought about by an improved office environment would improve worker morale, and probably other intangibles such as worker loyalty and commitment to the organization.

The seven building-in-use dimensions vary in their relative importance in predicting workers' environmental satisfaction. Their rank order is presented in table 6-2. Spatial Comfort is the most powerful predictor of people's work-station satisfaction, followed by Privacy and Lighting Comfort. Thermal Comfort is only a weak predictor of environmental satisfaction and psychological morale, but it is still important. The other three building-in-use dimensions are not statistically significant, although they do contribute somewhat to the 25 percent effect of environmental conditions discussed above. It seems ironic (and it is important to understand) that although other evidence shows that Air Quality, for example, is a dimension workers are very aware of in their office environment, it is not an environmental attribute that directly affects their morale.

The order of importance of the seven factors in predicting

TABLE 6-2
WORK-STATION SATISFACTION

Building-in-Use Dimension	Score*
Spatial Comfort	10.8
Privacy	6.3
Lighting Comfort	4.9
[Thermal Comfort]	

* T-test scores disproving the null hypothesis regarding the significance of this factor as a predictor of Satisfaction ($p > 0.001$, except for items in square brackets).

morale in the three buildings varies only slightly. Spatial Comfort and Privacy are important in all of them; Air Quality and Lighting Comfort vary somewhat in their influence as the third most important dimension. In interpreting what this means, it is important to remember that the fact that a building-in-use dimension is significant in predicting morale does not necessarily mean it is a positive experience for occupants. Lack of privacy could also be a predictor of dissatisfaction and low morale.

USING BUILDING-IN-USE SCORES TO PREDICT PRODUCTIVITY

Building-in-use scores are even better predictors of environmental workability than they are of environmental satisfaction. The term *workability* means the self-reported contribution the work space makes to individual workers' productivity. Productivity itself is not measured; rather, individuals assess whether their work space makes them more or less productive. This is a judgment they can make because their work is clerical, professional, or managerial—essentially white-collar. If their work were mechanical or manufacturing-related, man-machine systems theory would be a better predictor of individual productivity.

The seven building-in-use dimensions account for 40 per-

TABLE 6-3
ENVIRONMENTAL WORKABILITY OR WORKER PRODUCTIVITY

Building-in-Use Dimension	Score*
Spatial Comfort	14.5
Noise Control	8.6
Privacy	7.7
Air Quality	3.4
[Lighting Comfort]	

* T-test scores disproving the null hypothesis regarding the significance of this factor as a predictor of Satisfaction ($p > 0.001$, except for items in square brackets).

cent of environmental workability and people's own assessments of their productivity. This means that only 60 percent of the reasons people have for being able (or unable) to get their work done in an environment can be attributed to social and psychological factors, organizational factors, and situational effects. The building-in-use dimensions predict a greater percentage of environmental workability than they do of environmental satisfaction. The building-in-use dimensions that predict environmental workability, or workers' productivity, most effectively are listed in table 6-3.

This table is particularly interesting when compared to table 6-2. Spatial Comfort is the most powerful predictor of workability, as it is of satisfaction. However, Noise Control is in second place when it comes to predicting workability, whereas it does not figure at all prominently as a predictor of satisfaction. Privacy is a *less* powerful predictor of workability than it is of morale. This is interesting in view of the pervasive office-worker complaints about privacy (or lack of it) and noise problems. It is useful for managers to know that although Privacy is somewhat important to people getting their work done, it is more important in affecting their morale, and that noise complaints should be taken seriously because noise intrusion actually affects worker performance and is not just an issue of taste or comfort.

Air Quality is a weak predictor of workability, whereas it

does not feature at all as a predictor of satisfaction. This is perhaps contradictory to expectation, in that concerns about indoor air quality are widespread in modern office buildings. Also somewhat surprising is that Lighting Comfort is a stronger predictor of morale than it is of workability. People seem to have fewer expectations of lighting quality in terms of getting their work done than they do in terms of their environmental satisfaction.

In all three buildings, the building-in-use dimension scores account for more than 30 percent of occupants' judgments of workability. In the Eastern Building as much as 53 percent of the workability of the environment is attributable to the building-in-use dimensions. The rank ordering of the important predictors is quite similar in all the buildings.

That office environmental quality is an empirically important aspect of organizational functioning is demonstrated by the fact that the seven building-in-use dimensions account for 25 percent of people's environmental satisfaction and 40 percent of people's productivity. That Spatial Comfort and Privacy account for the major part of this percentage is critically important to managers, space planners, and office designers. This fact shows that it is the worker's microenvironment as created by his or her furniture layout and work-station design that has the greatest effect on both worker morale and productivity.

INTERVENTION FOR IMPROVEMENT

Although occupants' experience of Air Quality and Thermal Comfort are two of the most clear-cut dimensions of building-in-use performance, they are not the most important predictors of either workers' morale or their productivity. Spatial Comfort and Privacy are the major predictors of satisfaction and workability. What do Spatial Comfort and Privacy mean?

Spatial Comfort includes aspects of furniture arrangement and the spatial layout of the work station. It means access to files and document storage, convenience and suitability of boardrooms and meeting rooms, ergonomic suitability of the furniture, and the overall size of the work space. Privacy includes voice privacy, meaning that face-to-face conversations and telephone conversations are not to be overheard, and visual privacy, which is the degree of physical screening and control over personal accessibility. It would therefore appear that office workers could probably be rendered both more satisfied and more productive by improvements to the *spatial organization* of office-building interiors.

Spatial layout and furniture are cheaper to change than building-wide systems like air handling or lighting. It is reassuring to managers to know that workers' problems in contemporary offices are more a function of the microenvironment of the layout of the work station than of such macroenvironmental issues as air handling, which affect the entire building. Thus, a relatively small amount of microenvironmental change would appear to yield greater returns in terms of increased productivity than large-scale and expensive building improvements, all else being equal. These data are based on relatively satisfactory, well-functioning buildings. If a building has a serious air-quality problem, like the Terrasses de la Chaudiere, then repairing the mechanical systems is a priority.

In reviewing the meaning of these predictors, it is worth remembering that Lighting Comfort, which functions both as a predictor of morale and of worker performance, means quality of artificial light in the work space *and* quality of daylight in the work space. Unsuccessful building-in-use lighting means the workers feel there is glare from lights and/or windows, that there is inadequate or flickering light, and that there is little natural light or daylight in their work space. The other predictor that appears in these models is Noise Control, which means sound made by coworkers, both their voices and their use of equipment like keyboards and copiers.

Air Quality is a low-level predictor of environmental workability, whereas Thermal Comfort is a low-level predictor of worker satisfaction. This makes sense, in a way: Air Quality, which is judged from warmth, odors, and humidity, is a more physiologically sensitive dimension than Thermal Comfort, which is judged from coldness, temperature variation, and general temperature comfort. Generally, people feel they can work when they are slightly cold, even if it is not very agreeable or satisfying to do so. Most people feel they cannot get their work done if they are sleepy from high CO_2 levels or have headaches from stuffy air.

We have noted the interesting differences between what people like (i.e., what contributes to their satisfaction) in an office and what they need to get their work done (i.e., what contributes to their productivity). What other aspects of worker behavior in office buildings are affected by the building-in-use and can be predicted from the seven dimensions of the building-in-use?

USING BUILDING-IN-USE SCORES TO PREDICT HEALTH

One of the concerns of the modern-day office worker that was presented in chapter 3 is that their office is making them sick. The sealed windows, fluorescent lights, unpleasant odors, static electricity, and cathode ray tube (CRT) radiation that are typical of modern office conditions create an impression in workers that they are in an unhealthy environment.

To see how effectively the building-in-use dimensions predict building-related health complaints, the relationship between them was statistically examined. The building-in-use scores account for 21 percent of the variation in number of occupant health complaints, such as headaches, fatigue, colds and coughs, sore throats, neck and shoulder pain, faintness, and nausea. This means that just under 80 percent of health complaints are attributable to factors *other* than the seven building-in-use dimensions; these include other environmental factors, state of individual health, age, and sex, and individual variation in awareness of personal health symptoms. There are so many factors that can affect self-reported ill-health that the finding that 21 percent of workers' responses can be explained by building-in-use dimensions is surprisingly high. Table 6-4 shows the significant predictors of worker health. Out of the seven dimensions, Air Quality emerges as the strongest, followed by Lighting Comfort, and then by Thermal Comfort.

TABLE 6-4
BUILDING-RELATED HEALTH STATUS

Building-in-Use Dimension	**Score***
Air Quality	8.5
Thermal Comfort	5.2
Lighting Comfort	3.9

* T-test scores disproving the null hypothesis regarding the significance of this factor as a predictor of Satisfaction ($p>0.001$).

It is clear from this table that different building-in-use dimensions predict occupants' health status than those that predict occupants' morale and productivity. Air Quality is not a major factor in determining worker morale although it does appear to predict productivity; it is, however, critical to occupants' feelings of ill-health. As Air Quality is related to Thermal Comfort, it is not surprising that Thermal Comfort is also a predictor of ill-health. The building-in-use Noise dimensions, Privacy, and Spatial Comfort—so important to worker morale and performance—do not play a major role in occupants' feelings about their building-related health.

Lighting Comfort, while playing a minor role in predicting productivity, is more important in predicting people's feelings of ill-health and their morale. This tallies with what is known from other data about lighting in the work place, namely, that when asked to judge lighting quality, people often rate it as satisfactory but nonetheless report eye problems such as eyestrain and sore eyes. This apparent contradiction makes sense in the building-in-use context, where occupants' judgments of the interior office environment can clearly distinguish between conditions that affect their satisfaction, those that help or hinder them in doing their work, and those that affect their state of physical health.

From a manager's viewpoint, however, low morale, state of health, and ability to do work are all equally important in terms of how people perform in the work place. Both ill-health and dissatisfaction affect the productivity of office workers, the former through absenteeism and lost time on the job, and the latter through low morale and increased turnover. What are

TABLE 6-5
THE MOST IMPORTANT BUILDING-IN-USE DIMENSIONS

Productivity	Morale	Health
Spatial Comfort	Spatial Comfort	Air Quality
Noise Control	Privacy	Thermal Comfort
Privacy	Lighting Comfort	Lighting Comfort

the implications for management in being able to predict or explain these phenomena?

An understanding of what the building-in-use assessment is, how it can be rationalized in the context of other approaches to improving buildings for people, and an analysis and description of the seven building-in-use dimensions of environmental quality has yielded information about how to interpret the results of a building-in-use assessment in such a way as to make it most useful to decision makers. All seven of the building-in-use dimensions significantly affect environmental workability and worker productivity, followed by environmental satisfaction and worker morale, and then building-related ill-health. Table 6-5 summarizes the salient building-in-use dimensions in each case.

SUMMARY

There are seven dimensions or clusters of environmental ratings that represent building users' ways of experiencing the office environment. These represent alternative ways of assessing building quality, differing from the conventional diagnostic categories of thermal comfort, acoustics, indoor air quality, and lighting. The names of the seven building-in-use dimensions are: Air Quality, Noise Control, Thermal Comfort, Privacy, Lighting Comfort, Spatial Comfort, and Building Noise Control.

These seven dimensions represent seven sets of environmental ratings, which can be averaged to produce a score for each dimension. These scores are more indicative of how people judge environmental quality than single ratings, or than ratings clustered into the conventional diagnostic categories that correspond to instrument measurements of human comfort, but do not take psychological factors into consideration.

The scores for each dimension represent norms for the building-in-use database and yield the information that Privacy and Air Quality are normally at the low end of the scale of environmental quality, with Building Noise Control normally at the high end. However, it turns out that building noise is not a crucial element of workers' experience of environmental quality.

Spatial Comfort, Noise Control, and Privacy are the building-in-use dimensions that appear to be the most important contributors to workers' morale on the job and to the workability of their physical environment. Air Quality is less important to morale and worker performance than it is to workers' sense of their own health status in the buildings. Lighting Comfort is more of a predictor of morale and of worker health than it is of worker performance. The seven building-in-use performance dimensions play a significant functional role in office workers' productivity, morale, and health. They represent measurable aspects of environmental quality and are crucial for managers assessing buildings. Morale, health status, and productivity cannot be traded for one another; each of these, while related to the other two, is an equally important but separate component of worker psychology vis-à-vis the work environment.

Building assessment, to be effective, must be carried out in a social and political context: preferably the context of the organization that occupies and is responsible for the building in question. In the next chapter we discuss how to use the building-in-use dimensions and their scores: why organizations initiate building-in-use assessments, and how they use assessment information. We discuss who in the organization might carry out an assessment, what should be done with the findings, and ways in which the results of a building-in-use assessment improve office-building quality.

Chapter 7

IMPLEMENTING A BUILDING-IN-USE ASSESSMENT SYSTEM

Unless the results of a building assessment program are used by an organization as a basis for action, they represent nothing more than another numeric system for organizing information. Those who manage buildings and are in control of environmental quality through ownership or other means are best placed to make use of the assessment system.

Facilities management and building operation are areas that are traditionally short-changed by organizations. A recent article in the *Harvard Business Review* stated, "Failure to wring every benefit out of the most expensive capital asset most companies ever have would not be countenanced in any other aspect of corporate life."(Seiler 1984, 120) The concerns of building managers who feel that their responsibilities, being building-related, are not taken seriously by other executives, are reflected in this quote by a corporate property manager in a national newspaper: "We aren't taken seriously by senior management, . . . and a lot of companies put people in the job who aren't prepared. This is a serious business—millions of dollars are at stake." (*USA Today* 1986) If this is so, why are executives ignoring it? Possibly because traditionally there has been no simple "bottom line" way of proving that building improvement is worthwhile. The building-in-use assessment system offers a solution to this dilemma, and in this chapter we show step-by-step how such an assessment might be done.

Topics covered include *when* to measure environmental quality, i.e., how to recognize appropriate opportunities to get good results, *who* should initiate such an exercise, and *how* the outcome of such an exercise feeds back into the organization's decision-making processes. The discussion will cover ways of interpreting the findings of a building-in-use assessment, and how to make planning and design decisions based on the results.

USING ASSESSMENTS TO CONVINCE SENIOR MANAGEMENT

The building-in-use approach to assessing environmental quality is one way to persuade senior decision makers in organizations that policy decisions about their real estate inventory are as important as policy decisions about marketing, personnel, production, and sales. Its simplicity and ease of implementation make it both useful and easily understood. The system permits decision makers to attach a numerical quantity or score to the degree of environmental quality that exists in a given building. Once a number of buildings have been assessed, building owners and managers can use the system to make decisions about how and when to intervene to effect improvement. The fact that environmental quality in an organization's buildings can be quantified means that decisions about their building inventory can be "sold" to senior management in a straightforward and convincing way.

Traditionally, as pointed out in chapter 2, a building manager's decisions regarding environmental quality are *reactive* in that some clear problem has occurred in a building and money is needed to rectify it. For example, in warm weather, the chillers for a building's air conditioning system might turn out to be too small to enable effective cooling of the large amount of electronic equipment that has accumulated in the building. The organization occupying the building decides to solve the building manager's problem and that of the people who get overheated working at the electronic equipment by installing more and bigger chillers. But if workers still complain of overheating, noise, and lighting problems at VDT work stations, the facilities managers who recommended this solution look as if they have failed to solve the problem. Although they realize that a more comprehensive long-range plan for better VDT placement and management will probably pay off better in the long run, they have a hard time presenting this kind of long-range planning to senior management for approval and funding. Corporate executives often believe that single, one-shot solutions to solving building problems are the only type of solution worth considering. Facilities managers and planners can make good use of an inexpensive simple planning tool—the building-in-use assessment system—to illustrate that environmental change is an ongoing process of setting priorities and allocating resources.

This useful tool cannot be arbitrarily imposed on a situation. Building owners need to believe that they want the information it has to offer, and their employees have to be committed to careful and fair judgments of their work conditions. In chapter 1 we learned that some office-building problems have their origins in the commissioning and construction processes over which building owners and occupants have no control. However, organizational productivity and worker performance *do* lie within occupant control. The questions of which improvements, how much improvement, and how to establish priorities on improvements are those that are answered by the building-in-use assessment system. By using building users' own ratings of environmental conditions *as they experience them,* it is easy to pinpoint office buildings or parts of buildings that deviate from the norm for a corporation, a campus, or a single large building. Because certain dimensions of environmental improvement are more critical to worker satisfaction and productivity than others, and some of the deviations from the norms are greater than others, instant information is available on directions for building improvement.

THE FIRST STEP: ESTABLISHING PARAMETERS FOR THE BUILDING-IN-USE ASSESSMENT

Establishing the parameters of the database is the first step in setting up a building-in-use assessment. At this crucial stage organizational goals and constraints are built into the process that set the social and political context for the assessment.

Setting the parameters means responding to the question: "Environmental quality relative to what"? An organization that occupies a large office building of more than 95,000 square feet may decide to compare individual floors to the norms for the whole building, leading to actions for improvement that make all the floors in the building of comparable quality. An organization that owns several buildings of different sizes and in different cities, or parts of a city, may decide to establish norms based on all the employees' ratings regardless of the building they occupy, and to compare individual buildings to the norms for the organization. This allows the organization to implement company-wide norms or standards of environmental quality and to move personnel around without worrying about disparaging comparisons. Organizations that occupy several different types of building—for example, administrative office space, technical "knowledge worker" space, R&D facilities, or laboratory space—may wish to establish norms for each building type rather than for its work force as a whole, so that the norms reflect the limits on actual building use in each case, and managers can set improvement priorities accordingly. Companies that own or lease space in a building shared with other similar-sized organizations may decide to join forces to build up a database for the whole building so they can compare their own space to the norms for the building.

How parameters are set depends on what an organization wants to know about the quality of its work environment, and why. For the purposes of explaining the subsequent steps in building-in-use assessment, we will assume an organization that is interested in comparing one of its office buildings to the normative environmental quality of several of its other buildings, as the government did in the case of the Western, Central, and Eastern buildings.

THE NEXT STEPS: ASSESSING TARGET BUILDINGS

Once an office building has been identified as a target, workers are asked to identify on a scale of 1 to 5 their relative comfort with various environmental conditions. Some of the scales probe aspects of acoustics; others probe aspects of lighting and daylighting. Some ask about furniture design and spatial layout; others address thermal comfort and ventilation conditions. There are three or four scales for each dimension of environmental comfort, making a total of twenty-two ratings in all. People are asked to respond to all of them at one time, as quickly as possible, with no discussion among coworkers.

In small and medium-sized buildings, and buildings where people are very concerned or dissatisfied, all occupants can be asked for their ratings. In very large buildings, or for multi-building organizations, this may be too expensive, thus necessitating sampling of the population for practical reasons. When sampling, a carefully selected random sample of occupants from throughout the building is required to provide ratings that are representative of the whole building. In much the same way, Gallup samples approximately 1,200 people to elicit political opinion polls for the entire United States, reducing these to a few hundred for subgroup analysis. While sampling is practical, in reality, building occupants primed to provide ratings and interested in their environment may not accept that their opinions are being represented by only a few of their number, some of them perhaps people they do not work with and do not know. Organizationally and politically, it may make more sense to elicit ratings from everyone in the building, especially if a simple and effective way of doing so has been developed.

One building is assessed at a time. As responses come in, they are entered into a computerized database system that generates the normative scores not for each scale but for each of the seven building-in-use dimensions—the three or four scales that each dimension comprises. The scores for each building—the building-in-use profile—can be examined individually to give an idea of which dimensions are above average and which are below average in that building. Once the database has been established, individual buildings or their parts can be compared to the group; scores on the seven dimensions are computed for the test building and compared to the norms. The degree of *deviation* between the test building score and the baseline normative score is the second piece of useful information that is yielded by the assessment system: like the building-in-use profile scores, this is also a diagnostic indicator.

INTERPRETING BUILDING-IN-USE ASSESSMENT RESULTS

Table 7-1 shows the three office buildings described in chapter 4 compared to the norms for the group of office buildings to which they belong. Column 1 shows the *building-in-use profile:* the score received on each building-in-use dimension for the building. Column 2 shows the *norm* to which the building is being compared: the organizational baseline, which is the same for all the buildings. Column 3—the *building-in-use index*—shows the deviation of the individual building score from the group norm. Some of the numbers in this column are pluses, indicating that on those buildings the dimension is better than expected. Others are minuses, indicating that the building has a potential problem on this dimension. Zero on this index shows that the norm and the building score are the same: the building is normal on that dimension.

Looking simply at the building-in-use profiles without comparison to the baseline and the resulting index, we can infer that the Western Building is below average (i.e., below 3) in Air Quality, Noise Control, and Privacy, and is doing well on Building Noise Control and Lighting Comfort. Similarly, the Central Building's profile shows that it is lower than average on Air Quality, Thermal Comfort, and Privacy but needs no im-

TABLE 7-1
BUILDING-IN-USE ASSESSMENTS OF THE THREE BUILDINGS

	Building-in-Use Profile (Score)	**Organizational Baseline (Norm)**	**Building-in-Use Index (Score-Norm)**
Western Building			
Air Quality	2.2	**2.3**	−0.1
Thermal Comfort	2.9	**2.8**	+0.1
Noise Control	2.1	**2.9**	−0.8
Privacy	2.2	**2.3**	−0.1
Building Noise Control	3.8	**4.4**	−0.6
Lighting Comfort	3.3	**3.3**	0.0
Spatial Comfort	3.2	**3.3**	−0.1
Central Building (first survey)			
Air Quality	2.3	**2.3**	0.0
Thermal Comfort	2.6	**2.8**	−0.2
Noise Control	3.3	**2.9**	+0.4
Privacy	2.4	**2.3**	+0.1
Building Noise Control	4.6	**4.4**	+0.2
Lighting Comfort	3.2	**3.3**	−0.1
Spatial Comfort	3.3	**3.3**	0.0
Eastern Building (Tower 1)			
Air Quality	2.2	**2.3**	−0.1
Thermal Comfort	2.8	**2.8**	0.0
Noise Control	2.7	**2.9**	−0.2
Privacy	2.2	**2.3**	−0.1
Building Noise Control	4.0	**4.4**	−0.4
Lighting Comfort	3.7	**3.3**	+0.4
Spatial Comfort	3.3	**3.3**	0.0

provement on the other four building-in-use dimensions, appearing exceptional, in fact, on Building Noise Control. The Eastern Building scores are also poor in Air Quality, Noise Control, and Privacy, and are better than average on Building Noise Control and Lighting Comfort.

Although the profiles provide an interesting summary of building-in-use performance in each case, showing at a glance the strengths and weaknesses of each building, the profile scores alone are an insufficient basis for action for change. Until the scores have been compared to organizational norms, we do not know in what ways these buildings are or are not "normal," that is, meet normative expectations. The decision to take action for improvement is a function of the building-in-use index rather than the profile.

A look at the building-in-use index shows that the only dimension that needs action for improvement in the Western Building is Noise Control (−0.8). The other lower-than-average scores, although low, are not different from the expected score, or norm. The high-scoring Building Noise Control is also lower than expected, showing a minus on the index that is almost as large as that for Noise Control. The Central Building's index shows that it is not lower than expected on any of the building-in-use dimensions and is in fact significantly better than the norm on Noise Control (+0.4), *not* Building Noise Control, as the profile suggests. The Eastern Building is unexpectedly low on Building Noise Control (−0.4) and unexpectedly high on Lighting Comfort (+0.4). Its other dimensions are all normal.

Table 7.2 summarizes the different types of information from the building-in-use profile and index for each of the three buildings. Although they do not say the same thing, both are correct. Whereas the profiles are descriptive in the information they provide, the index priorities are a basis for action for change. The profile describes building performance, and the index shows where the building's performance differs significantly from its expected performance on the seven dimensions. The index does not contradict the profile but gives the same information passed through another "filter" (i.e., norms, expectations) so that priorities for action can be determined.

TABLE 7-2
COMPARISON BETWEEN PROFILE AND INDEX PRIORITIES

		Profile	**Index**
Western Building	*Problems:*	Air Quality Noise Control Privacy	Noise Control
	Successes:	Building Noise Control Lighting Comfort	—
Central Building	*Problems:*	Air Quality Thermal Comfort Privacy	—
	Successes:	Building Noise Control	Noise Control
Eastern Building	*Problems:*	Air Quality Noise Control Privacy	Building Noise Control
	Successes:	Building Noise Lighting Comfort	Lighting Comfort

It is important to remember in the implementation of a normative system to differentiate between what is *expected* and what is *acceptable*. Because the low profile scores on Air Quality and Privacy in these buildings are expected and not abnormal, this does not make them acceptable. The information from the index that Noise Control is abnormal requires immediate attention; the information from the database norms that Air Quality scores universally low is an issue requiring long-term attention from the organization. The criteria for acceptability on each dimension of the building-in-use must be developed by the organization doing the assessment as part of its goals, mission, and policies regarding building quality.

DETERMINING THE URGENCY OF ACTION FOR CHANGE

This comparison of meaning between the building-in-use profile, which tells some of the story of office quality, and the building-in-use index, which tells a lot more of the story of office quality, demonstrates the principles of building-in-use assessment. The profile score on each building-in-use dimension is a diagnostic descriptor: it is a general indicator of the environmental quality in that building. The building-in-use index is more than descriptive: it is a diagnostic indicator of how to plan environmental improvements. The third stage of assessment is to determine how large an index difference warrants how much follow-up action by managers.

A brief statistical calculation will enable users of the system to ascertain the significance of the degree of deviation of the building score from the norm (i.e., the index).* This depends in part on the number of respondents providing ratings for each building, and in part on the value of the norm itself. At least two levels or degrees of deviation of the score from the norm can be readily identified. Where the index is zero, there is no deviation and the building is normal. At the first level of deviation, the dimension that does not score as expected should be tagged or identified for further investigation. At the second and more extreme level of deviation, the building-in-use is performing noticeably better or worse than expected. In the case of the latter, at least, a response is urgent.

With the additional statistical calculation, each index incorporates two items of information for each building-in-use dimension. The dimension is either performing better or worse than expected (or normally), and it is either slightly or seriously better or worse than normatively expected (unless it is normal). This breakdown is shown in figure 7-1. Each dimension in each

---

* Application of the Central Limit Theorem and the use of z-scores to calculate confidence levels from the sampling distribution.

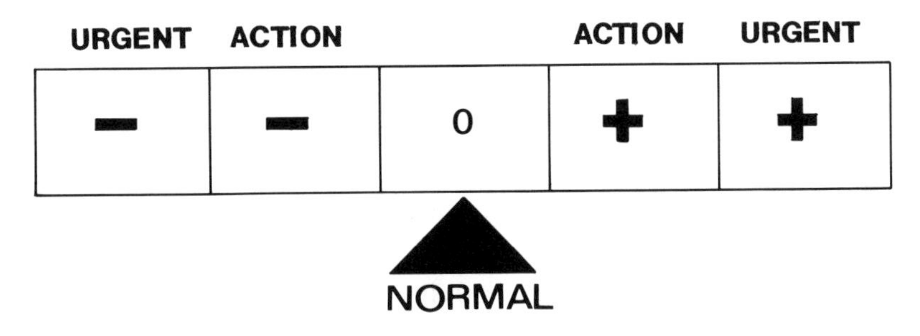

7-1. Action for change.

building can be located in one of the five cells on this matrix, each of which represents a guide to action for change.

Table 7-3 illustrates one last important factor in interpreting building-in-use results. Chapter 6 identified which building-in-use dimensions made major contributions to employee morale and to workers' performance and health. In addition to the meaning of the building-in-use profile scores, the comparison to norms, and the usefulness of the building-in-use index, the numeric findings displayed in table 7-1 must also be interpreted in light of how we know each building-in-use dimension affects workers' behavior. Although all seven building-in-use dimensions have some effect on these three areas of occupant behavior, certain of them are more influential than others.

In interpreting the meaning of the index for each dimen-

TABLE 7-3
THE MOST IMPORTANT BUILDING-IN-USE DIMENSIONS

Productivity	Morale	Health
Spatial Comfort	Spatial Comfort	Air Quality
Noise Control	Privacy	Thermal Comfort
Privacy	Lighting Comfort	Lighting Comfort

sion, and the cell into which each score falls, it is important to remember which dimensions are important to worker behavior. Table 7-3 shows that Spatial Comfort, Privacy, and Lighting Comfort are probably the most important building-in-use dimensions overall, and therefore even small deviations in the building-in-use index on these dimensions should be attended to immediately.

For example, although both Noise Control and Building Noise Control are poor in the Western Building, the finding on Noise Control is more important because this dimension affects worker performance more prominently than Building Noise. In the Central Building, Noise Control rates excellently on the index, showing, in fact, that noise levels are not uncomfortable for workers in this building. Knowing what we do about this building, further investigation would confirm that this does not necessarily mean there is no noise. It means that Central Building employees do not require protection from noise to get their work done, and that they therefore do not judge high noise levels adversely. The Eastern Building index suggests that both Lighting Comfort and Building Noise Control require further investigation. Although these are both urgent according to the index, Lighting Comfort is a more important factor than Building Noise Control in both worker satisfaction and worker health, so its poor index is potentially a more serious and pressing problem.

TYPES OF FOLLOW-UP ACTION

There are two categories of follow-up action to the findings from a building-in-use assessment. One is the implementation of *additional testing* of a building-in-use dimension to acquire more information and a more precise definition of the problem. The other is *action for change,* such as building alterations, changes in maintenance and operation practices, and the devising of a plan of actions for improvement.

What action may be taken as a result of a building-in-use

assessment is a policy decision related to organizational goals. The second category of follow-up—action for change—involves setting priorities on change by allocating resources to tasks and activities. Primary resources in this case are operating dollars, staff and consultant time, and management's attention and effort. These resources are allocated in a time frame that corresponds to the urgency classifications of figure 7-1: short-range or immediate action on "urgent"; mid-range action on "action"; and long-range action on those changes that will pay off in the long run, that will improve the operation of the organization.

The first category of follow-up—additional testing—is undertaken if a problem needs more precise definition before it can be solved. Additional testing includes the use of measuring instruments and data collection along the lines of those described in chapter 5 for lighting, ventilation, thermal comfort, and acoustics. Because instrument measurements are often lengthy and complex to undertake, their use as a follow-up to specifically targeted areas of poor building performance is a way of making them more efficient and useful in defining (and therefore solving) problems.

Follow-up measurements provide additional information on building occupants and their tasks as well as on building performance. The source of an acoustical problem, for example, may lie in the high confidentiality demands of the task. An acoustical problem may also be caused by hard, sound-reflective surfaces in the office space. Where a low score is received for Air Quality and Thermal Comfort, rigorous structural and mechanical testing is important as worker health may be at risk. These items, if poorly rated, pose a more immediate threat to individual health than poor acoustics, which may result in a decline in worker performance but are unlikely to cause worker absenteeism unless the effects are stressful over time.

For each building-in-use dimension, a low score means different kinds of follow-up testing. In the case of a low Privacy index, complaints about privacy may or may not lend themselves to follow-up with instrument measurements. Privacy problems may be acoustic or visual or may indicate a problem of environmental control. They may have to do with a specific problem, such as high privacy requirements for a given task, or they may simply be a generalized expression of malaise in the office or work group. More precise definition of a privacy problem is required through interviews with office staff, task analysis, and careful study of the furniture arrangements and spatial layout of the area in question. It is a good idea to remember that privacy complaints are somewhat endemic in modern office space: everyone complains about lack of privacy even if other environmental elements are the real problem. It is unusual to find Privacy rated well in a building where no other dimensions are evaluated well. Poor ratings on other building-in-use dimensions tend to accompany poor Privacy ratings. Interpretation of a building's index score on Privacy must be made with caution; other dimensions must be investigated before a solution is prescribed.

Follow-up testing of Spatial Comfort means examining an office space in more detail, through observation, interviews, focused groups, and other social survey techniques. The use of instruments to assess details of Spatial Comfort performance requires measuring with a ruler or tape measure the dimensions that are relevant to ergonomic comfort, such as height and width of work surface, height of chair seat, depth of seat pan, placement of chair back relative to seat height, and relative distance from the worker of tools and equipment such as telephones, keyboards, and reference materials. Another aspect of Spatial Comfort is overall size of the office space, which may warrant careful attention if much of the space is shared. Furniture design and placement also affects lighting, acoustics, and thermal comfort; these need to be examined to determine whether a poor Spatial Comfort rating is related to one of the other dimensions.

Where the dimension receiving a low index is Air Quality or Thermal Comfort, additional testing is strongly recommended. Poor Air Quality may be a health hazard. Poor Thermal Comfort is unusual and therefore worth following up. The long-term implications of a low Air Quality rating require building performance analysis as well as more information from users, which might take the form of individual interviews with workers in the poorly rated area. Air Quality testing includes carbon dioxide and carbon monoxide level checks and possibly monitoring, testing air samples, and installing monitoring instruments for such known toxic agents as ozone and formaldehyde, both of which are routinely found in office environments. Building-in-use Air Quality includes air movement, warmth, and humidity. Therefore, the r.h. also needs to be tested and air circulation monitored to determine how well air is moving through the space.

A poor Thermal Comfort rating means uncomfortable coldness and drafts. Ambient temperature, air speed, and dry and wet bulb readings should be taken. Other tests include detailed examination of the air distribution system, determination of whether the sensors are accurate, whether the ducts are clear and airflow unobstructed, whether the fans are working properly, and, in the case of a VAV system, whether the VAV boxes are in good working order and their thermostats are functioning.

If a Noise Control or Building Noise Control problem is identified, instrument testing will help determine whether its source is in physical elements of the environment or in worker needs and expectations. Noise transmission through and reflection off various physical elements in the office can be measured; sound levels, and whether noise is continuous or intermittent, familiar or strange, caused by people or machinery, are all topics for further analysis. However, noise problems in the office often have more to do with the discrepancies between acoustic conditions and task requirements than with measurable amounts of noise. It is therefore important, when following up on low noise ratings, to analyze the acoustical requirements of workers' tasks before introducing instrument measurements of levels of sound.

Lighting Comfort complaints indicate opportunities to measure amount and direction of light on the task and on the surround, and to calculate reflectance of surfaces and contrast conditions. Lighting problems are often difficult to identify because of the complex interaction between human vision and measurable illumination conditions. Other important aspects of lighting problems include the colors and luminance of work-station surfaces (including carpets, walls, partitions, and furniture), sources of glare from lights or windows, and the visual health of the person working under these conditions.

Before a blanket program of tests is implemented, remember that each test yields enough information to pose better and more refined questions; it does not necessarily yield answers. As the amount of information accumulates, assigning priorities to its degree of importance becomes more and more difficult. Some tests are more accurate than others; many are cumbersome and expensive, requiring careful evaluation before they are implemented to determine the payoff from doing them.

However carefully it is planned, additional testing cannot do more than yield more data that translates into information about the problem. Acquiring more information will not, by itself, solve a building problem; it will help define a basis for action that will solve the problem. Thus, the two categories of follow-up defined above—gathering more information and taking action for change—are also a two-step follow-up to building-in-use assessment. If needed, more information will provide a better definition of the problem, but the ultimate aim of follow-up is to solve the problem, and that means taking action to effect improvement.

POLICY DECISIONS AND CORPORATE GOALS

The amount and type of follow-up action for poorly rated building conditions depend in part on the general goals and objectives of the user organization, in part on the degree of corporate commitment to building quality, and in part on the expert knowledge of and resources available to managers. Building-in-use assessment is a way of furthering the objectives of senior management in terms of assessing environmental quality, quantifying it to enable the information to be used as a basis for action and long-range planning, and providing a context in which buildings or parts of buildings can be assessed. Once the assessment has occurred, the system is operating, and numbers are in place, senior management must judge the findings. In this section, some critical policy issues posed by building-in-use assessment are identified.

The first of these is deciding on the seriousness of the degree of departure of the building-in-use profile score from the norm for each dimension. Even with the additional statistical information regarding the confidence one can have in the degree of deviation of a building-in-use score from its norm, *selection* of the acceptable degree of deviation of each score from the norm is still a policy decision.

The other critical policy area is the norm itself. The fact that a corporation finds its baseline norm is below 3, and even around 2, as are Air Quality and Privacy in the database described here, is a call for another approach to follow-up action, different from those described above. The corporation must decide how high or low a norm is *acceptable* for each building-in-use dimension of its building stock.

Raising a low normative score requires a long-range planning approach to action for change. This may take various forms. One is to increase consumer pressure on building designers to use ventilation technology creatively and to solve the problem of unacceptable indoor air quality. Another is to take remedial steps in each building to improve the normative rating. These include improvements to building ventilation systems and to the temperature of the air, its humidity, and its odors, as these are all part of building-in-use Air Quality. People's attitudes may also change if they receive more information from managers and have more opportunities to give feedback to management on their experiences of the building environment.

A third approach is to "market" office space more effectively. In a small office building in Massachusetts, the builder-developer told the occupants when they moved in that a special air handling system had been installed which guaranteed a complete recycling of all the air every hour. The feature was written up in a trade journal as the solution to afternoon drowsiness, and the article was photocopied and passed out to the building's tenants. The workers in this building feel very secure about the quality of the indoor air as a result, and if

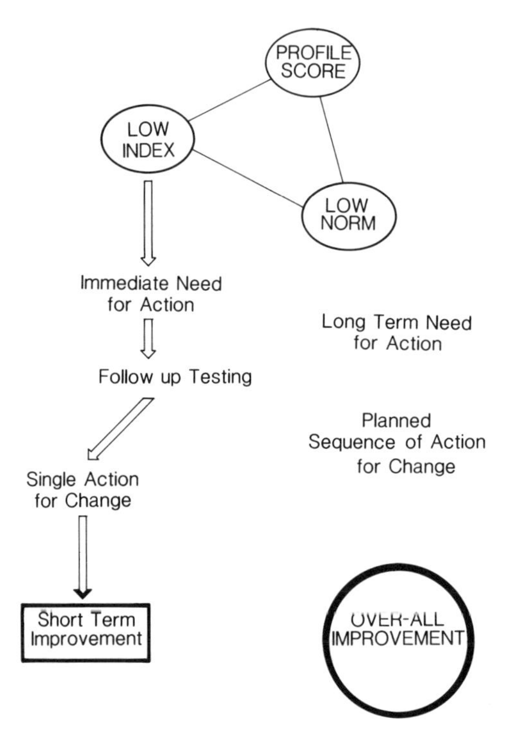

7-2. Directions for action based on score, norm, and index.

asked would no doubt rate ventilation above 3. There are many possible solutions to low building-in-use norms!

Solving a low norm problem means accumulating more knowledge and a better technology; it is a long-term goal of building improvement. Figure 7-2 shows how the different approaches to follow-up relate to the time frames for action for building change.

This figure indicates that the *index,* the difference between the score and the norm, yields information for immediate application; this information pertains to the specific building. The index provides a basis for managers to plan, to set priorities, to solve building problems, and to effect short-term improvement. The *norm* provides information about existing standards and user expectations of the building stock; it is a basis for long-term planning and the allocation of resources to generic building improvement.

Corporate standards regarding which normative scores are acceptable for which building-in-use dimensions, and planning guidelines that set policy regarding how much follow-up is warranted by the degree of deviation of each building-in-use index, must be established as a function of the goals, objectives, and desired image of the organization. For example, a government agency may want the norm on every dimension of building performance to be 3 or over so that it can make a public statement to that effect. On a consistently poor dimension such as Air Quality, management may choose 2.5 as the Air Quality standard for its buildings as this may be cheaper to maintain. Air Quality may become a political issue in buildings where management is satisfied with 2 but the union wants at least 4. Whereas low norms on all dimensions may satisfy a public agency as long as none of the scores sinks below 2, private or corporate building owners with a greater concern for corporate image and building quality may be motivated to maintain building-in-use norms of around 4. Building owners may also market good-quality office space by saying that when in use, Lighting Comfort, for example, has performed at 4, or that all building-in-use dimensions are at 3 or better.

Scores and norms that indicate building excellence also warrant follow-up action; not necessarily for improvement, but for a better understanding of what makes a good building. Although a score of 4 on Building Noise Control is not excellent because it is close to the norm, a score of 4 on Noise Control approaches an excellent score. An addition to the wisdom of building managers and operators is guaranteed by an investigation of the reasons that so little discomfort is caused by noise in a building where it scores 4. This is the type of knowledge that will be applied in the long term to new building design and large-scale renovation. Criteria for environmental quality can be synthesized from acquiring knowledge of what works best in existing buildings. The past decade or two of building research has failed to close the feedback loop between building designers (architects, engineers) and information from building evaluation. Knowledge of what makes a good building will allow the formulation of criteria for good performance that can be applied to the design and specification process as a way of ensuring environmental quality in a building while it is still in the planning stages.

SOME KNOTTY PROBLEMS OF BUILDING-IN-USE ASSESSMENT

Although there may be good, rational reasons for an organization to assess building quality, there are some procedural difficulties that are social or organizational in nature. These may result in no more than tentative approval for an assessment program with no funds or follow-up forthcoming from senior management.

The first of these procedural difficulties is the near-universal concern with the notion of evaluation. The idea of evaluating something is never popular; the very word *evaluation* often makes people nervous. Architects are concerned that their building design will be criticized; workers are concerned that their job performance will be found wanting; building managers and operators are afraid that they will hear more than they want to know about the problems of the buildings for which they are responsible. In this book, the term *assessment* has been used rather than *evaluation,* specifically to avoid the value-judgment connotation. The building-in-use assessment is an assessment of relative environmental quality that can be used as a tool by facilities managers and others concerned with operating buildings. But building evaluation, albeit for specific and practical purposes of effective maintenance and smooth operation, is the goal of this system. It is therefore important, when undertaking a building-in-use assessment, to avoid the term *evaluation* as much as possible, to quiet the doubts and anxieties the idea arouses, and to stress the non-value-judgment connotation and practical applications of the building-in-use assessment system.

The second procedural difficulty in implementing building

assessment is in the use of questionnaire surveys to elicit ratings from building users. Many organizations are leery of questionnaires and user surveys. For many managers, giving out questionnaires to workers means that workers lose time they should be spending on the job, are asked to think about things that are not part of their job description, and get the idea that solutions to all their problems are just around the corner. When solutions are not forthcoming, as is usually the case, the manager has to deal with dissatisfaction, low morale, and even mutinous behavior. A not altogether unfounded suspicion exists that giving people questionnaires is not ultimately useful, requires a lot of time and money, and is potentially more trouble than it is worth.

The fact that the building-in-use assessment system is based on workers' ratings of environmental conditions does not mean that a traditional questionnaire survey needs to be carried out. In fact, in an automated office, electronic methods of eliciting ratings can be used. The database is derived from the ratings themselves, not from a questionnaire. People are asked to assign a number between 1 and 5 to a scale of comfort on an environmental condition. They can circle numbers on a piece of paper, telephone numbers into a central office or even a computer, or record their ratings through their personal computer in the same way that they use electronic mail and message services.

Building occupants do not need more than ten or fifteen minutes to rate twenty-two scales, and they should all be done at the same time. Once the ratings are in, workers should not need to spend time on them again, at least for a while. In carrying out a building-in-use assessment, make sure that workers have a clear understanding of why they are making these decisions, what the numbers they are circling mean, and how the data will be used. Emphasize to them that the *individual* response is neither identifiable nor interesting: the assessment system is a function of group responses and group norms, and workers may be "grouped" in a number of different ways to make the data useful (for example, by floor, by work group, by sex, by degree of work-space enclosure). Explain the process sufficiently so that workers understand enough about the assessment to take their responses seriously and to think about their answers. Note that complicated questions about "difficult" topics such as job satisfaction, organizational change, and promotion prospects are unnecessary for the purposes of accumulating an assessment database, and are not part of the survey.

The third procedural difficulty is at the level of meaning and interpretation: using a normative system carries the risk of excluding important information that is not normative. Applying building-in-use assessment norms as standards against which to judge environmental conditions carries with it the risk of ignoring information from the extreme ends of the range of occupant responses. People who react to a building condition such as polluted indoor air with serious health problems may be in the minority and statistically may represent the extreme end of a response distribution curve. Nevertheless, these workers' complaints are proportionally very serious, and it is important for a building operator to be able to flag health-related building problems early and pay attention to them even if they are not initially uncovered by a standardized survey. One sick worker taking days off to recover may be as serious a problem to a manager—who sees the effect on the group's productivity—as a group of uncomfortable employees who complain a lot but are not indisposed enough to stay away from work. A woman having pregnancy problems, or a few workers unknowingly breathing toxic substances, are as serious a problem for an organization as a whole floor of dissatisfied complainers. Both are serious problems and require management action.

Figure 7-3 diagrams the relationship between the *scale of user complaint* and the *scale of seriousness of the problem,* showing when managers should act on user information. This diagram illustrates how managers should apply the results of the building-in-use assessment to their decision making about the office environment. As the figure demonstrates, it is as important for a building manager or decision maker to flag a small number of people (or even one person) with a health complaint as it is to flag a large number of people reporting a reduced ability to work or environmental dissatisfaction.

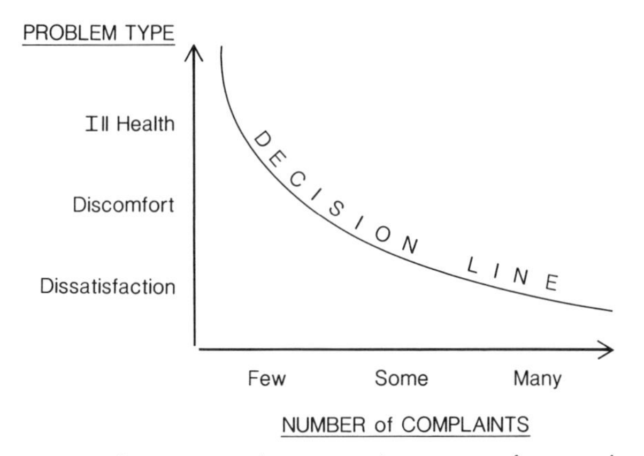

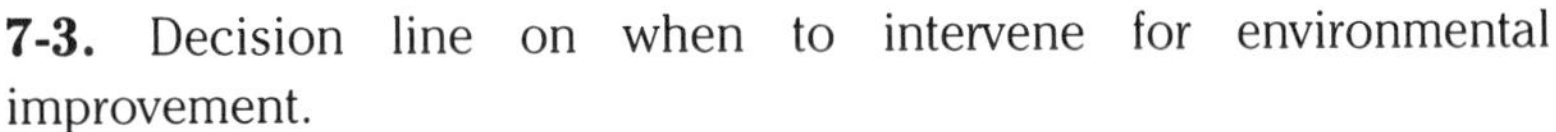
7-3. Decision line on when to intervene for environmental improvement.

SUMMARY

Once a building-in-use database has been established and performance assessments of individual buildings have been executed, some buildings do well, while others do not. Privacy and Spatial Comfort may score low; Noise Control and Thermal Comfort may score high. The owning and/or occupying organization has several courses of action open to it as a result of the knowledge it has acquired.

At the first level of information is the building-in-use profile—the scores between 1 and 5 that the building received on each building-in-use dimension. This is primarily descriptive information. Managers can use graphic techniques to overlay plans of the buildings with their profile scores on the dimensions. Such displays will show at a glance which buildings or parts of buildings are above average (3) on which dimension and which areas fall below average.

Decisions regarding follow-up action are based on the index, or the difference between the building-in-use profile scores and the organization's norms. These decisions include whether to gather more information for a more precise definition of the problem; whether instrument measurements are required to define possible causes of the problem; whether and how soon to budget resources to solve the problem, either by intervening with the users (e.g., changing their tasks, moving them, replacing supervisors or workers) or with the building (repainting walls, adding desk lamps, installing a sound-masking system). Priorities on taking such actions are set by the urgency of the index, and, in the long term, by where the norm itself falls on the scale of 1 to 5. Decisions regarding follow-up action are made with reference to corporate standards and planning guidelines that must be set at a policy level. Follow-up actions may be in the form of a one-year or a five-year plan; they may be part of routine operation and preventive maintenance tasks; they may involve structural and procedural changes within the organization.

Important cautions in implementing a building-in-use system include believing that *evaluation* is not a dirty word; avoiding comprehensive occupant questionnaire surveys; and not excluding extreme degrees of complaint by relying on normative assessments of the environment.

Once the scale, type, amount, location, and timing of building improvements are established, effecting environmental change is often a simple proposition. Because the building-in-use performance assessment system ensures that the improvement will be effective not only in reducing complaints but also in improving worker performance, it should not be difficult to sell the idea to senior executives. Who can quarrel with the small investment in the building that is needed if improved organizational productivity is the outcome? While there may be some costs involved in setting up the building-in-use system, the payoff—a guarantee that real building improvement will result from the expenditure of building improvement dollars—is hard to resist. If it is done properly, this is a no-lose proposition. At worst, the building improvements implemented with the advice of the building-in-use assessment will be no less effective than current ways of making decisions about buildings. At best, people will receive improvements to their environment that will really enable them to do their work better.

Guidelines for good interior office design are presented in the next chapter. They provide some ideas about specific environmental changes that result from the building-in-use assessment system follow-up and that improve environmental quality. They are based on the state of our knowledge about human behavior in office buildings to date. They provide a basic inventory of environmental quality criteria for application to new and renovated buildings. Their application to office renovation and design will result in the improvement of the state of the art of the modern office building.

Chapter 8

WHAT MAKES A GOOD BUILDING? GUIDELINES FOR ENVIRONMENTAL IMPROVEMENT

The desire to improve environmental quality in offices is part of the job definition of building managers, part of the expectation of office workers, and part of the commitment of most corporate employers. Information regarding the design of good office space is becoming increasingly available: some is relevant to architects and engineers, some to interior designers and space planners, and some to building operation and maintenance staff. This information is often contradictory and often speculative; much of it is incomplete. However, by applying as much of this knowledge as possible to the design or redesign of office interiors, improvement in environmental quality is guaranteed.

PURPOSE OF THE GUIDELINES

Guidelines based on the state of our knowledge about human behavior in office buildings synthesize information and make it useful. Guidelines provide a basic inventory of environmental quality criteria to apply to new design and to the improvement of the building-in-use dimensions in below-normal environments. Environmental design guidelines in a checklist form are a useful tool for building planning and space design.

As the previous chapters have shown, each building-in-use dimension represents the interactive effects of a number of ambient environmental conditions. Effective design guidelines must therefore incorporate knowledge from the various traditional building science disciplines and must re-present this information interactively. In this chapter the topics covered by each set of guidelines are introduced and defined before the guidelines themselves are listed. The guidelines include:

- background information on each topic;
- conventional wisdom and current practice;
- reference to standards of health and comfort where they exist;
- recommendations regarding the best way of handling specific environmental situations;
- performance criteria for environmental quality and quality control.

These guidelines are intended to guide action for environmental improvement. In each building situation, the priorities for environmental improvement will have been set as a result of the building-in-use assessment. A poor score or rating on an environmental condition in a building (or part of one) is elicited from the building assessment system to indicate the need for change to and improvement of that condition. Even without problem analysis, however, there is still no reason why lighting, acoustics, and ventilation/thermal conditions should not be adjusted and improved wherever possible in offices so that each group work space and individual work station is optimally designed for its user within the constraints of the building itself and the limitations of our knowledge.

The foremost issues in designing an office interior are the distinction between the open-plan and cellular-plan office, and determining which is best for you. Modern planners take the position that this is not an "either/or" question. There is no need to polarize a concept that is neither open nor enclosed space; rather, a continuum exists along which a range of enclosure choices is available. The question is, how open is open plan? How many people have to be grouped in a single space before it is no longer a shared office but an open office? In some organizations, space managers have decided that "the open plan" is less suitable to their employees than single-person enclosed offices. In others, clerical staff are in open offices, while professional and managerial staff are in enclosed single offices. Technical staff are often in enclosed shared offices. This kind of spatial decision making often has little to do with efficient use of the building for workers' tasks and more to do with differentiating the rank and status of occupants.

Some organizational analysts recommend space planning that is neither open nor enclosed but allows people choice in the kinds of spaces they need depending on the tasks they have to perform (e.g., Allen 1977; Stone and Luchetti 1985). The gist of this approach to office planning is to dispense with individual work spaces and encourage teamwork, meetings, individual tasks requiring concentration, and file storage to occur in different, related office areas that are accessible to all workers depending on their needs.

It is commonly held that staff with managerial responsibility need individual enclosure for privacy reasons, whereas—depending on the tasks they perform—other office workers can manage without full enclosure and indeed may benefit from the increased communication and visibility of the open plan. However, workers' tasks may vary, so that clerical people working on personnel files need full enclosure for privacy, whereas technical/professional people are often away from their desks and have little need of an enclosed work space. The specific requirements of workers' tasks need to be examined, as do the structure and processes of the groups within the organization, before planning decisions about enclosure are made final. Amount and type of work-station enclosure affect the users' experience of ambient environmental conditions including lighting, acoustics, and thermal comfort, and therefore affect the ways in which office quality can be improved.

These space design and planning guidelines do not address specific degrees of enclosure or openness. Where appropriate, specific reference is made to open/enclosed differences. For example, acoustical problems in an open situation are different from those that arise in a situation where each worker is enclosed, and they are handled environmentally in different ways.

The guidelines are practical because they aim for maximum improvement without large costs. They represent a set of criteria against which degree of environmental quality can be determined, for example as a quality control checklist. They are designed to correct errors of judgment or carelessness in spatial layout and physical design. They assume conventional building technology in a building that is either in the design and construction stages, or that is already built and being renovated. In modern new office buildings, large front-end expenditures are often more likely to go on the façade, the entry and atrium, the windowed elevators on the exterior wall, or innovative solar panels, than on workers' furniture, interior color and design, and innovative ventilation technology.

Not that the former items are irrelevant to workers: on the contrary, a nice-looking, bright, and well-kept office building provides strong incentives to workers to stay put, and even attracts new ones. However, the fees paid to big-name architects and engineers to produce ultra-chic-looking buildings are often disproportionate to the environmental quality of their interiors. The guidelines that follow provide a quick and concise overview of practical ways of ensuring and improving office environmental quality. They are organized into the seven building-in-use categories as follows:

- Spatial Comfort and Privacy
- Lighting Comfort
- Noise and Building Noise Control
- Air Quality
- Thermal Comfort

SPATIAL COMFORT AND PRIVACY

The guiding principle of office interior design in contemporary buildings is flexibility. In most new office buildings, mechanical and electrical systems are specified by the design engineers to accommodate a range of possible interior layouts and uses. These include open-plan offices, cellular-plan offices, and everything in between. Even when the building is purpose-built for a work group or an organization, the exact layout of people at work inside the building is usually not known when crucial mechanical and electrical decisions have to be made. As a result, most systems are supposed to accommodate all possible interior layouts, ranging from no people at all to crowded conditions.

In most modern offices, interior spatial layouts are expected to change at a rate of 30 to 70 percent per year. As a result, most office spatial layouts are in a state of flux. An office building could change 100 percent in less than two years in the way its interior space is used. The building shell is expected to accommodate this rate of change in the interior. Although some of these changes are major and some are quite small scale, office furniture and spatial layout must be easy to alter and ultimately flexible to lend themselves to change and to make best use of the fixed aspects of the building, the systems that are in place before people move in.

A major challenge posed to interior space planners when designing a work space is the protection of workers' privacy. Earlier chapters indicated that complaints about lack of privacy dominate users' judgments of environmental quality, especially where the floor is laid out in open plan. However, privacy can mean many different things. The dilemmas of privacy are that as a worker one needs to be seen, but not watched; one needs to communicate, but not to be overheard; one may work on a computer that can be accessed by other workers, but one does not like one's terminal screen read over one's shoulder. And finally, and most difficult for a space designer, providing individual privacy may isolate the worker from the group and consequently reduce the work group's effectiveness, so that a design that meets the individual's need for privacy may not be the best overall design for a group work space.

Another challenging problem for office planners is that not all office workers are constantly at their desks. More and more offices have a large number of people who move around the building, meeting with different people or using different pieces of equipment as their job requires. In many organizations, there are employees who are often not in the building at all. Their work may require them to be on the road, as in sales, or supervising other offices, or in manufacturing plants located elsewhere. In most conventionally designed office buildings, however, it is impossible *not* to continue delivering heat, light, and air to empty office space.

The guidelines listed below address some of the more important aspects of flexibility in office planning and furniture layout. They are not intended to be an exhaustive space planning manual; rather, they are intended to show how other aspects of environmental quality hinge on crucial aspects of furniture and spatial layout.

WORK-GROUP SPACE DESIGN

Group Size

Plan space for work groups only after determining the size of the work group, how many rooms or enclosed spaces it either requires or can manage with, the privacy requirements of its various tasks, and its relationship to other work groups, including degree of separation and sharing of equipment and facilities. These are all a function of the task and organization of the work group, which have to be understood before a good work space can be planned. A good spatial layout is based on communications analysis and studies of paper-flow, filing, and accessing stored information (Harris et al. 1981).

Participatory Planning

In planning a group's work space, integrate the constraints of the physical building and its systems with the task and activity requirements of the work group and with the budgetary and other conditions imposed by the organization. Do not allow space planners to be placed in a reactive role (i.e., reacting to the demands of individual workers).

A participatory planning process *can* be used to plan a work group's new space, as long as it is carefully managed and controlled. Base a participatory planning process on more than canvassing individual occupants. Individual workers who are asked about their spatial preferences impose individual requirements on the space planner with little reference to the group task and with no reference to the building's own constraints and limitations.

Create a participatory planning process that encourages workers to respond to models or plans of alternative spatial layouts that meet the criteria of the work group and of the organization before finalizing the design of the new space.

Absent Workers

Provide separate environmental zones for work-groups that do not spend much of their work time in the building. This enables heat, light, and air handling to be switched off when their work space is not in use. This can result in considerable energy savings to the organization, depending on the proportion of people who spend time outside the building.

Storage

Handle storage in a manner similar to privacy: i.e., different tasks have different requirements. Subsume individual preferences in the interest of the work group. Individuals will fill as much paper storage space as they receive, even though this may not always be in the interest of or necessary to the work of the group.

It is generally accepted that no office worker can actively use more than two linear feet of files at one time. Make group storage accessible and generous so that people do not feel the necessity for large areas of personal file storage. Control excessive paper and file storage by individuals with changes to the organization of work (if necessary), with accessible group or centralized file storage, and with a firmly enforced policy about using it.

Make larger storage areas that do not need to be part of the individual work station available to the work group. Like conference rooms and copying machines, design group storage as a shared resource.

Circulation

Narrow-floor offices, or offices designed around atria, have circulation dictated by available floor space. In deep-floor buildings, circulation has to be designed into the work-group arrangements. Group and organize workers according to tasks and task requirements, and let circulation move around them, helping to define them. In planning circulation through an office floor, take advantage of window views and other orientation cues.

Signage and Orientation

The people who work in a building are not its only users. In some buildings, especially government offices, members of the general public are regular visitors and need to be accommodated. Although their needs are not as pressing as those of the people who are in the building all day every day, they have certain requirements of an office environment that affect their experience of that building. Of these, the most important is probably signage and orientation.

Avoid large bland open spaces with repeated visual elements that can be disorienting from the inside, as people try to find their way around a floor. Use color and finish materials to communicate directional and orientational information; design adequate and legible signage; and provide plenty of visual access to window views and daylight to improve wayfinding and orientation in the interior of office floors.

Visitor Space

Mark waiting areas clearly for visitors to office buildings, so that the offices and waiting rooms they seek can be found without disruption to office work. Such areas, the population and use of which will fluctuate depending on what the visitors are waiting for, how often they come, and how long they stay, should be supplied with their own exhaust fans, and if a large area is involved, with a dedicated air-handling system. These areas do not require careful lighting and work-station layout, although colors, especially colored signage, may be important in visitors' areas.

ACOUSTIC PARTITIONS OR SCREENS

Space Planning

Using acoustical screens to re-create an illusion of single-person enclosed offices is a poor application of the open-plan concept. Heavy use of acoustic partitions in an open plan generates lighting, ventilation, and orientation problems. Workers may hear a noise or a voice and be unable to identify its source; it creates a nuisance effect that distracts from work.

Use acoustical screens as they are intended to be used: to provide an extra modicum of acoustical privacy between two workers whose privacy is compromised by the requirements of their tasks and by the absence of other acoustical elements.

Visual Screening

Visual screening is often given as a reason for invoking heavy use of acoustical partitions in an open plan. However, there are other ways of effecting visual screening in a deep-floor open-plan layout that can be employed before acoustical screens are introduced.

To meet selected visual privacy requirements for workers who wish to exercise their option of withdrawing from the group for certain tasks, orient desks and chairs away from public view, provide plants with foliage for screening, place tall furniture judiciously, and provide alternative work spaces for special tasks (e.g., quiet rooms for report writing, or meeting rooms for interviews).

Acoustical partitions should be used for increasing (not ensuring) acoustic privacy between an individual worker and a noise source. For workers who need visual and acoustic privacy for most of their daily tasks, full enclosure using floor-to-ceiling walls can be created either for their own work space or as a shared resource for the work group.

Lighting

Where screens or partitions are 60 inches or higher (the minimum height desirable for acoustic privacy), three- or four-sided enclosure of each desk by screens prevents workers from seeing around them. They have access to less natural light and window views than they would in a more open layout. Some of the light from an overhead source is absorbed in the cavities created by colored screens, and this results in less light on the task (the work surface).

Supervisory staff enclosed by screens for acoustic privacy have to get up from their seats to supervise the workers around them adequately. Glass or Plexiglas can be used to enclose one side of a work space to solve this problem, but they do not provide effective visual screening and can result in a fishbowl effect, in spite of possible lighting advantages. Glass and related materials reflect sound and can increase noise in an enclosed space.

Acoustical Screening

Create opportunities for private conversations, confidential paperwork, closed meetings, or personal telephone conversations in the office space, but not necessarily at every desk. Design office space so that such opportunities can be shared in the same way that equipment such as copiers, terminals, and conference rooms are shared.

Some tasks, such as personnel counseling, require 100 percent visual and acoustic privacy because of their confidentiality requirements, and in these cases, privacy should be provided. However, most jobs require only occasional or intermittent privacy, in which case a nonpermanent private space, with shared use as needed, is adequate.

Enclosure

Most people prefer to have the option of not being fully available to others in the office at all times. Workers agree that it is hard to get work done with the constant interruptions of office conversation, meetings, and telephone calls. They feel that an open plan encourages the illusion of interpersonal availability.

Use screens and other controls judiciously to modify such behavior without resorting to full enclosure for every desk. The acoustical partitions in most furniture systems do not provide total acoustic privacy; they reduce the efficiency of the illumination; they impede air circulation in an open-plan space; and on a large floor they produce a labyrinth effect that can be very disheartening (Passini 1984).

LIGHTING COMFORT

In recent years, work in office buildings, like work in industrial settings, has generally been illuminated by high levels of overhead lighting standardized throughout the work space. Lighting, usually integrated into the ceiling system, provides a uniform level of overhead light for all office tasks. If adjustments are necessary for reduced lighting, or even increased lighting in some of the offices, they are often made later, after workers have moved into the building. Such lighting systems are operated in zones, with centralized switching.

Until the "energy crisis" raised awareness of buildings' energy consumption, using natural light to light office tasks was the exception rather than the rule. Even today windows are generally discounted by designers as sources of light, either because they generate glare or because the amount of light they supply is unpredictable owing to varying weather conditions. In fact, it is customary for lighting engineers to specify lighting for office interiors that ignores the amount of light provided by windows. They do so by specifying lighting that provides enough light for all-dark conditions, regardless of window design, natural light, and the amount of nighttime use expected in the building. Lighting designers have also tended to increase the amount of artificial light emitted at the perimeter, close to windows, to minimize differences in the visual field between the bright window and the dark surround of interior walls.

Conventional wisdom has held that the more invariable (constant) and intense the light, the better it is for workers. Recent thinking indicates that this is not the case, but many conventional work environments continue to be designed along these lines. Many people dislike fluorescent lighting, without really knowing why. The "humming" sound of fluorescent ballasts can be irritating. Some people in offices have their fixtures "delamped" or keep them turned off. Some people place incandescent lamps on their desks. Some people believe that light from fluorescent fixtures makes their skin and coloring appear strange; some people believe the subliminal frequency at which fluorescent lights flicker exerts a noxious effect on their vision; and others believe fluorescent light causes cancer. None of these convictions has been empirically supported by research studies. More importantly, though, none has been categorically repudiated.

For these and other reasons, designers have come to recognize the attractiveness of daylighting to office workers and have invented various methods of introducing daylighting deeper into the interior of each floor, including the use of light shelves and interior courts. Solar technology is being developed to bring sunlight into basement rooms and workrooms below ground as well as into dark stairwells and building interiors.

Research has shown that office occupants in interior spaces often overestimate the amount of daylight that reaches their desk from the perimeter windows, provided they can see a window. Thus, with careful spatial layout and attention given to the lighting of each office task, a little window light goes a long way.

INDIVIDUAL ADJUSTMENT

It is important to analyze the visual demands of workers' tasks before designing the lighting of their work spaces. Artificial lighting should be designed to complement and add to the daylighting in a space rather than the other way around. Because people's lighting requirements are individualized and may change as their tasks change during the day, or over several weeks or even years, design a lighting system that enables occupants to adjust as many aspects of their lighting conditions as possible without having a negative impact on the other workers in a space.

TASK-AMBIENT LIGHTING

High levels of constant overhead lighting are not really necessary, especially if task lighting is available. Offices designed to be lit by 700–900 lux from overhead fixtures with no task lighting are typical but not well designed. Lighting of this type is unlikely to be found in an expensive building. High overhead light levels lend an institutional quality to the interior space that is reminiscent of schools and warehouses.

With switchable task lighting provided for every occupant, the overhead or background lighting in an open office space can be as low as 300 lux, and in Europe it often is. Specify a task-ambient system using low levels of background light and individual or zoned switchable task lamps, if possible. Although such a system is expensive to install, it is energy efficient over the life of the building and helps create a high-quality environmental experience for occupants.

WARM LIGHTING

More and more office buildings are using so-called warm-white fluorescents instead of conventional, white-light fluorescent lamps. The warm-white fixture supplies light covering more of the natural spectrum than the cool white light, which is concentrated at the blue end of the spectrum, and it seems to be preferred by workers both for office tasks and for the way it makes them look.

Although they are more expensive to buy than white-light fluorescents, warm-white fluorescents are no more expensive to operate (unlike incandescent lamps), so install them throughout the office.

HID ILLUMINATION

Although energy efficient, High-Intensity Discharge (HID) lighting should be used cautiously in office buildings as they are often disliked by workers for their poor color rendition and excessive brightness. HID lighting is best used in public areas, like atria or high-ceilinged corridors, and in areas like warehouses or garages where workers are not doing fine visual tasks. Poor color quality makes it undesirable for cafeteria or kitchen use.

HID lamps are not easily switched on and off and controls are usually zoned. They can be used in office interiors for background or ambient lighting, where workers have warm-white or incandescent task lighting in a task-ambient system. Metal Halide HID lights have better color rendition than sodium or mercury vapor lights, but a combination of types of HID lighting gives the best color rendition.

CONTRAST CONDITIONS

The color and reflectance of floors, work surfaces, walls and partitions, and dominant pieces of furniture have a major effect on the quality of good lighting. Dark desk tops are not desirable for paperwork if there is major contrast between the dark desk and the white paper. The task should be lighter than the desk surrounding it in a ratio of not more than 3:1 (Harris et al. 1981).

Design task lighting to light the task from above and to the side rather than from directly in front of and above the task, to avoid contrast problems from veiling reflections. Similarly, avoid blotters or desk pads on desk tops covered with clear plastic or glass, especially where the main light source is overhead: they reflect light into the eye of the user and create glare. Glass-covered or highly polished desks in some organizations are unfortunately a symbol of status. They are only acceptable if the desk user does almost no work at the desk. For example, managers and supervisors may be in meetings or on the telephone most of the day, and if they are not writing or reading at their desk, the reflected glare will not be so harmful.

DAYLIGHTING

Accompany the use of daylighting in buildings by effective solar controls. Where windows admit light from large areas of sky, specify that they must be able to be covered when the sun is bright.

Daylighting design includes reflectors for redirecting daylight into interior zones, light shelves, and light-colored ceilings and walls. Heat gain is controlled by shading devices and reduced-light-transmission glazing. Photocells are available for installation into light fixtures. These link to dimmers that reduce the amount of artificial light emitted when daylight conditions are bright.

To prevent excessive contrast between wall and window, which creates glare, do not allow the ratio of window-to-wall area to exceed 3:1. The use of design features such as overhangs, shading, light shelves, and window shape and orientation (as well as the climate itself) can dramatically affect the amount of bright light admitted by a window. This ratio should therefore be considered as a guideline rather than a fixed rule.

VDT users often suffer from window glare, either because a window is behind them and light is reflected in the screen, or because the window is behind the screen and provides a surround many times brighter than the screen. In order not to deprive VDT users of natural light altogether, use clerestory windows in VDT office spaces.

COLORS

Colors in the surrounding space affect the efficiency and effectiveness of a light source. Dark colors absorb light, for example, and reduce light-source efficiency, whereas light colors reflect it, and increase its efficiency. Avoid large contrast differences in adjacent colors in the visual environment, as this can cause glare, which will affect workers' visual comfort.

On the other hand, avoid a single neutral color, a solution that is often used in office buildings to reduce questions of taste in color to a minimum and for ease of maintenance. A single neutral color is visually bland and generates a general feeling of discomfort from lack of stimulation (Lam 1977).

Use colors of varying hue and saturation to generate variety, but of the same brightness to avoid glare. Illuminate office areas that have massive or dark furniture more brightly than open or light-colored areas are illuminated.

Most window light is reflected from the ceiling, and least from the floor, so as long as ceiling and walls are light-colored and clean, floor colors can be darker and still allow natural light to be effective.

When asked about lighting, most people will say their preference is for incandescent light. This is because it favors the warm (red and yellow) end of the spectrum. Consider incandescent lights for desk lamps for certain tasks (e.g., those requiring more interpersonal interaction than writing and desk work), as incandescent light is not an efficient light source and is expensive to use throughout a building.

GLARE FROM FIXTURES

Make sure that the brightness contrast between visual tasks close to the eye and distant surfaces that are also in the line of sight of the worker are not too large. Do not exceed a ratio of 1:5

between the brightness of the task and the brightness of the surround. Larger-contrast ratios produce glare, which generates visual discomfort. Avoid light sources in the line of vision of the worker, as these can be an additional source of discomfort. Recess overhead lights and/or shield them by louvers or lenses from the line of sight of the *seated* worker: many coffered ceiling systems are designed to shield the eyes of a standing person, but this is not useful for most office work.

MAINTENANCE

Lighting maintenance and repair is an important contributor to illumination efficiency. Develop a maintenance schedule that requires acrylic lenses to be cleaned regularly. To maintain lighting efficiency, replace aging fluorescent and HID lamps regularly as they emit less light as they age.

Avoid a policy of "spot relamping" as this will result in uneven and inefficient light distribution. Replace all lights in the building on a regular basis to maintain even light emission and reduce the occurrence of humming ballasts.

VDT LIGHTING

VDT equipment has special lighting considerations. If windows are nearby, be certain that they are clerestory windows or are shaded to avoid reflections in the screen and to reduce glare. Use only subdued overhead light sources, out of the workers' line of sight, and possibly shaded. If possible, background lighting should be indirect (e.g., a wall wash) rather than overhead. Make good-quality adjustable task lighting available to VDT users.

Provide opportunities for VDT users to change their focal length with a long-range view, preferably through a window. (This is difficult to do because of glare considerations.) Experiments are still going on to determine how best to light VDTs, and consensus on an optimum lighting condition has not yet been reached.

NOISE AND BUILDING NOISE CONTROL

The objectives of good acoustic design are to enhance wanted sounds and control unwanted sounds (noise). People prefer to work in an environment that is quiet but not totally free of sound. People use sound for orientation and awareness and also for masking individual voices and conversations. People in offices also need to communicate easily, both with each other and on the telephone, without the strain of shouting to be heard or the stress of feeling that all conversation is overheard. Different office tasks and activities vary widely in their acoustical requirements. Most offices are designed to standard acoustic specifications that do not respond to this wide variation in requirements.

The principles of good acoustics include: a relatively quiet background; freedom from intruding noise; adequate loudness of the sounds we wish to hear; proper distribution of sound to all listeners; and protection of those who are not meant to be listeners from hearing that same sound (Wetherill 1985). An adequately quiet background is provided by the building enclosure and its exclusion of outdoor sound, and controls on the noise from building systems (such as lighting and ventilation). Adequate loudness of desired sounds (e.g., your own conversation) and proper distribution of sound in a space depend on the shape of the room and the type of materials used as finishes. Some materials absorb sound and some reflect it; some materials and structural systems transmit sound. Adequate protection of office workers from *unwanted* sound (noise) in their offices depends on spacing and furniture layout, on the finishing materials, and on the design and the sound-transmission rating of various types of partitions.

One of the difficult aspects of designing a good-quality acoustic environment is defining the different types of noise and how to manage them. The qualities of steady background noise are different from those of intermittent intrusive noises, and both types of noise can be disturbing for different reasons. These reasons have more to do with how expected or predictable the noise is, and if it is understandable and meaningful, than with its loudness or proximity to the hearer. Steady loud sound may cause fatigue but is not always perceived as immediately stressful the way intermittent loud sounds are.

FINISHES

Sound levels in open-plan offices are largely controlled by sound-absorbent finishes on the ceiling, floor, and sometimes walls. Use sound-absorbent finishes such as glass fiberboard to control reflected sound levels. Use sound-blocking materials that are dense and impervious as a barrier between rooms or work stations. Specific acoustical requirements for a task may require some work space enclosure, such as the individual desks of telemarketing operators.

Do not use carpet on walls or office partitions to control sound, because, although it is soft and thus does not reflect sound, it is not very sound absorbent. If you use acoustic tile on ceilings to absorb sound, or at least to prevent it from being reflected back into the space, combine it with other elements of acoustical design, such as desk orientation, acoustic partitions, and absorption on sound-reflecting surfaces.

SPACING

Control sound traveling through open-plan offices by *distance, spacing,* and *orientation* of workers and equipment. Voices and other sounds carry in an open space when they are reflected off hard flat surfaces such as windows, some ceilings, and the flat acrylic lenses over light fixtures. Cover or soften flat hard surfaces that reflect sound by recessing light fixtures, treating ceilings and if necessary walls, and covering windows.

People needing privacy should orient themselves away from windows and other hard surfaces when they talk on the telephone. Offices that are completely or partially enclosed for acoustic privacy can transmit sound via connecting windows—a common problem in modern buildings with continuous perimeter glazing. Specific treatment (sound-absorbing drapes, a wall, or a partition) is necessary at the window where it connects adjacent offices.

SOUND MASKING

The installation of a background sound-masking system should only be considered after you have acoustically planned and treated your open office. A uniform bland sound such as a "whoosh" that is too quiet to be intrusive and is maintained in the correct auditory range and frequency to mask speech is often used to protect voice privacy. Sound that carries information, such as music, is not effective for sound masking.

Do not use sound masking alone to create good acoustical quality in offices. To be effective, sound masking should be used along with other acoustical design elements. If unwanted sound is not reduced in other ways, the sound-masking noise just adds to existing noise levels and can create an uncomfortably high level of background noise. Sound masking does not function as *the* solution to the problem of poor acoustics: use it as one in a series of steps to improve office acoustics.

Many people confuse the "whoosh" of a sound-masking system with the sound generated by an office air handling system and are reassured that their office is receiving a continuous supply of air. Although people dislike sound levels that are intrusively high, they are equally uncomfortable in a too-quiet space with no background or masking sound. The intrusiveness of unwanted sound is in part a function of its meaning and the information it carries. Many office workers object to Muzak systems, for example, because they like other types of music and because they cannot "tune it out."

AIR HANDLING

In a sealed building, occupants like to hear some sound from the air handling system, as this assures them that it is on. However, if it is a variable air volume system, some people become alarmed when the system cycles off. As a result, engineers often prefer air handling systems to be silent all the time. Unfortunately, this sometimes convinces users (especially those seated in poorly ventilated areas) that they are not receiving any air at all.

In a large number of offices, the background "whoosh" of air handling systems is the only sound masking available, and the desirability of knowing that air is being delivered outweighs the intrusiveness of the background noise. Although acousticians counsel against using air handling systems for sound masking, consider this as a solution to low-background-sound situations. People's dislike of the sound-masking "whoosh" coupled with their desire to know that there is ongoing ventilation may make this a more effective sound-masking option.

As an alternative to sound masking, consider providing nonintrusive background noise that is pleasing but not informational. Although office workers complain about noise, they do not want a completely silent environment. Consider running water as a background sound for an open office. Introduce a small fountain or waterfall—as in the NMB Bank in Amsterdam, Holland—as part of the interior design to make an attractive contribution to a well-designed office space and to mask sound effectively and unobtrusively.

ACOUSTIC PRIVACY

Enclosing offices with walls or partitions does not guarantee a solution to acoustic problems. To be effective in reducing voice transmission between workers in an open plan, specify acoustic partitions that are at least 60 inches high and extend close to the ground.

Sound is transmitted both through partitions and over them. To exclude sound effectively between enclosed offices, extend walls through the ceiling to the concrete slab, baffle vents and ducts, and do not allow dropped or hollow floors, which are often suggested for cabling purposes. The costs involved in constructing this kind of acoustic insulation are high and often excessive for the degree of acoustic privacy required for most office tasks. However, this is often done in lawyers' offices, for example, where confidentiality is important. If office work groups require complete acoustic privacy for some of their tasks, provide a shared multiuser conference room that has been fully acoustically insulated for this purpose, to be used on an as-needed basis. Acoustic insulation, after all, is a matter of degree.

Provide enclosed offices or conference rooms with walls that have Sound Transmission (STC) ratings of 45–55 decibels for personnel with management or hiring responsibilities who conduct client meetings or have other requirements for confidentiality. An acoustically insulated conference room may not even be required on every floor except in office buildings where corporate secrets are being traded or government policing or auditing agencies have offices.

EQUIPMENT-GENERATED NOISE

A common acoustic problem in the open-plan office is office equipment that generates noise. This noise does not have to be loud for workers to find it disturbing; persistent noise is also tiring and stressful because it contributes to an overall high level of sound in an enclosed space. Even the clatter of VDT keyboards can create a noise problem when there are many together in one room. Typewriters, copiers and printers are well-known noise generators. If such equipment does not come with sound covers (which should be used), order sound covers or have them made.

Enclose larger machines, such as mainframe computers, in their own rooms. Locate noisy equipment outside office areas or where the noise is least intrusive to workers.

AIR QUALITY

Ventilation and air quality are frequently criticized and elicit some of the more vigorous criticism from office building occupants. In most modern office buildings, HVAC engineers design air distribution systems that incorporate heating, cooling, and humidification for the whole building, regardless of the range and variety of conditions that may exist in the building. Each air distribution system in a building has a zone, or area of the building to which the fans deliver air, and from which they extract it. Often, the larger the building, the more zones it has; but frequently the very largest buildings have only three or four. Some have separate fan systems on each floor, but often these feed into and out of a single large air handling system in the basement or in a "penthouse" on the roof. Such large systems are economical to design and run, but the principle of large systems for large buildings poses air quality problems.

For example, if one small area of the building is introducing a contaminant or pollutant into the air, such as a chemical from printing or laboratory work, or something from outside in the street, or smoke from a group of heavy cigarette and cigar smokers, their air is incorporated into the system and redistributed throughout the building. Engineers often surmise that the contaminant is so diluted by the volume of air into which it is being mixed that it can do no damage, but this may not be the case. Although there may be no direct threat to health, there is likely to be discomfort among people inhaling the contaminant.

If a pollutant is considered toxic in high concentrations (often the only standard that exists for air pollutants), it may also be toxic in low concentrations over a longer period of time. The relationship between toxicity, time, and amount ingested is unknown for most of the chemicals found in the air of sealed buildings. The volume of air in which the contaminant is mixed varies according to the amount of air that is being recycled in the building. Most air distribution systems work on a principle of mixing a small amount of fresh air (i.e., air from outdoors)—often as little as 10 percent—with the air already in the building. For each 10 percent of fresh air introduced, 10 percent of the old air is exhausted out of the building or assumed lost through cracks (exfiltration), and the rest is mixed together and recycled. Because fresh air often needs to be heated or conditioned, the more fresh air is introduced into a building by the mechanical systems, the more costly the building is to run. Often, building operators will rely on "infiltration," or the inadvertent admission of outdoor air into the building, and make use of "economizer cycles" which introduce no fresh air at all. As a result, the old air, with all its pollutants, is reused and some buildings may take considerable time for a complete change of indoor air.

Workers are mistrustful of air in sealed buildings, and not without reason. Numerous studies indicate that even though people know outdoor air in many cities is polluted, they would feel better if they could open their windows. Although modern windows are designed to admit a very small amount of all the air needed in an office building, the fact of being able to open them makes people feel safer about the air they are breathing. In fact, in some of the more recent buildings, operable windows are offered as a luxury item.

STANDARDS

ASHRAE specifies that an air handling system should introduce between 5 and 25 cfm (cubic feet per minute) of fresh air per building occupant per hour.

In a typical office environment, ensure that the relative humidity (r.h.) of the building does not drop below 25 percent. The building enclosure should be designed to tolerate higher levels of humidity, up to 40 percent. An r.h. higher than this will encourage the growth of bacteria, pathogens, and molds in the ventilation systems and may cause condensation in the building envelope that will lead to long-term damage. An r.h. lower than 25 percent causes people to experience eye irritation, sore throats, chapped lips, respiratory problems, and static electricity.

AIR CIRCULATION

Tall furniture, such as filing cabinets and acoustic screens, especially those that extend to the floor, even if they are no higher than 60 inches, may impede air circulation even if an adequate *amount* of air is being delivered to the space. In many office buildings, air is delivered from the ceiling through diffusers, the majority of which are located at the perimeter of the floor. Air that has circulated through the office space is usually returned through ducts or plenums from ceiling vents in the interior of the floor.

All equipment and furniture on a floor ventilated in this way represent impediments to the circulation of air to the people who are on that floor. Because air handling for most office floors is designed for an open and unobstructed space, walls and partitions built to enclose work stations or meeting rooms can also prevent the efficient flow of air. Make sure that the path of air delivered at the perimeter is essentially unimpeded, so that it can move successfully through the interior and is drawn down to a seated worker's nose level before being returned through the ceiling.

If acoustic partitions are used on a floor, make sure they clear the floor at their base to a height of 1 to 2 inches in order to allow air flow. If walls or full-height partitions are used for enclosure, ensure that within each enclosed space there is at least one supply vent (diffuser) and at least one return or exhaust vent. If most of the air is being supplied at the perimeter of a floor, provide an adequate number of return vents in the interior of the floor to ensure proper air flow.

BALANCING

Most air handling systems are designed on the basis of a totally open floor with no walls, furniture, or people. Any changes to that floor plan, especially the addition of floor-to-ceiling walls, means the system must be checked to ensure that it still delivers and returns air in the way it was designed to do. Once an office floor has been filled up with furniture and people, balance the air handling system to ensure that the right amounts of air are being delivered and returned. All too often, the system is inadequately balanced or not balanced at all. Rebalance every air handling system after a year or two, depending on the building's "churn rate," or rate of moving work groups around, and after changes of occupancy.

In a surprisingly large number of modern office buildings, either no humidification system has been installed, or the system is not functioning properly. In northern climates with cold dry winters, humidification is as important in the winter as dehumidification is in the hot humid summers. The right amount of humidity in the air can make a significant difference to office workers' experience of the environment. Require the design engineer to specify a functioning humidification system.

Air removal is as important as air supply, and often in meeting rooms where several people congregate, the amount of air exchange in the room is inadequate. In such cases, provide low-noise air conditioners or exhaust fans in these rooms to ensure that adequate air moves through the room when the door is closed. Provide a switch for the extra equipment so that the extra air circulation is only used when the room is in use.

SOURCES OF POLLUTION

Provide all special office equipment, such as copying machines, printing and reproduction equipment, microfiche readers and printers, and shredders, with a way of removing fumes and dust from the workers' air and preventing these pollutants from being recycled into the building's air distribution system. One way to do this is to provide a dedicated exhaust system in areas where such equipment is located. Alternatively, solve this problem through the judicious location of such equipment in the building. For example, place copiers near the doors to toilets so that fumes are exhausted by the toilet exhausts, especially in buildings with low-pressure air handling systems where much of the air circulating through the building is exhausted in this manner.

Dust and particulates are often generated in office buildings owing to the high level of paper use and movement. If this is a problem in your building, here are a few simple measures to take. Ensure that the building's humidity is maintained at around 40 percent to reduce the amount of dust flying around. Supply the building's fan rooms with adequate and frequently changed air filters to control paper dust and some of the heavier particulates.

Cigarette smoke is a major source of respirable particulates and other pollutants in indoor air. More and more buildings are banning smoking. Policies on controlling smoking in buildings abound, as this is more of a behavioral problem than a design problem. If cigarette smoking is confined to specific rooms or areas of the building, make sure that these spaces are ventilated directly outside and that the air is not recycled, even partly, for reuse in the building's interior. Even if smokers' spaces have their own air handling systems, migration of pollutants across systems is common, and smoke cannot be kept completely out of other areas of the building. If smoking is not controlled in the building, increase ventilation rates to remove smoke as quickly as possible.

In buildings where workers' complaints about air pollution are more pronounced than the norm, or where people are sicker than might be expected, check building interiors for contaminants from unsuspected sources. Check for carbon monoxide leaks from underground garages and delivery docks; check vicinity of air intakes for sources of pollution; check kitchen and toilet exhausts or laboratory exhausts that may leak back into the building or cross over into other air handling systems in the building; and check for new furniture, carpets, and drapes that may give off formaldehyde, ozone, and other chemicals that cause discomfort and nausea.

If necessary, follow the example of the Environmental Defense Fund's national headquarters in New York: buy furniture and fittings made of nontoxic materials like solid wood, wool, and stone, and use an organic nontoxic wall paint, if you can afford it!

NEGATIVE IONS

Research has shown that many aspects of modern office buildings, including ducted air conditioning, static electricity, cigarette smoke and other contaminants, and high density of people, contribute to the depletion of ions from the air. Use ion-generating machines in some work spaces to generate negative ions for office interiors, as these seem to affect office workers better than positive ions.

Although not all people are ion-sensitive, as many as three-quarters of a given population may be so. The beneficial effect of adding negative ions is more pronounced for women and for workers at VDTs. The beneficial effects are also more pronounced at night (Hawkins 1984).

ENERGY CONSERVATION AND AIR QUALITY

"Flush" sealed buildings overnight and, especially with new buildings, over the weekend, to allow completely fresh air to greet occupants in the mornings and to remove contaminants while workers are not in the building. This practice is sometimes stopped in the interests of conserving energy, but the discomfort reported by workers in the mornings when they arrive (especially after weekends) and can smell the stuffiness and leftover pollutants from the previous day suggests that in the long run it is not worth the dollars saved. If you do not "flush out" the building, you will probably have to deliver more air while the building is in use during the day to replenish the stale air.

Energy-saving measures in some buildings have resulted in less fresh air being introduced into the building during the day. Make regular checks of building operating practices to ensure that the minimum recommended fresh air standard is observed (at least 6 cfm per person; preferably 15 or 20 cfm per person; 25 cfm if the building allows smoking).

Control the static charge resulting from dry air, manmade fibers of carpets and drapes, and electronic equipment—especially VDTs that have a static charge on the screen surface—through higher humidity, static-control floor covering, and negative-ion generators.

THERMAL COMFORT

Although a considerable amount of research has been done on the topic of people's thermal comfort under various working conditions, it does not seem to be the most critical element of environmental quality in office buildings. This is probably because extremes of heat and cold are unlikely to exist in office buildings, and people are already by and large working in environments that have been designed to provide an envelope of comfort within which individual needs may vary but individual discomfort is manageable. Moreover, people's thermal comfort depends only partly on the ambient condition that is supplied by the building, and partly on conditions they select themselves. For example, people may vary their clothing to adapt to temperature, and in some jobs they may vary the amount of physical activity in which they engage, or at least space it out to prevent themselves from getting either too hot or too cold. Most offices provide thermostats that people may alter to suit themselves or the needs of the work group.

In fact, the availability of thermostats is often a "placebo," having the effect of satisfying workers' thermal comfort needs psychologically, without having any real effect on ambient temperatures. In some buildings they are designed not to be adjusted by anyone except those building management staff members who carry instruments that will open up the box. In other buildings the thermostats are linked to the adjacent space and do not affect the zone in which those who adjust it are working. In some office buildings, thermostat settings adjustments have generated so much conflict and bad feeling that the thermostats have been disconnected and HVAC systems relying on sensors in the ducts and ceiling substituted.

Organizations such as ASHRAE have well-developed sets of thermal comfort standards for interior spaces. A formula has been worked out to provide a "Comfort Envelope" within which thermal comfort measurements indicate that some 80 percent of users will be comfortable. The measurements from which the thermal comfort rating can be computed include ambient temperature, air speed, relative humidity, and clothing and activity levels.

Thermal comfort sensations are closely linked to air quality perceptions in users, especially where warmth is concerned. If occupants feel the air is too warm, they also experience it as stuffy and possibly polluted. Cooler temperatures feel fresher, but occupants have a low tolerance for temperature fluctuation. Research has indicated that a constant temperature (cool or warm) is usually unremarkable to office workers; problems arise when there are unpredictable temperature shifts. People at work dislike drafts, sudden cold air, rapid changes, and fluctuation. The conventional temperature provided in office buildings is 70–75°F, although this was lowered to 68–73° during the energy crisis.

COOLER TEMPERATURES

Because heat is generated in all buildings by people, equipment, and lights, HVAC systems have to work harder at cooling a building than heating it. As a result, contrary to what one might expect, cooling offices is more energy-consuming than warming them. HVAC systems in large buildings are typically designed to cool interiors, even in winter. Thus, an overall lowering of the indoor temperature standards could result in a greater consumption of energy. Moreover, a number of HVAC systems adjust the amount of air they deliver according to the temperature required, as indicated by the setting on the thermostat. Thus, a requirement for cooler air may result in *less* air being delivered to the office space.

Recent research indicates that people at work feel less fatigue at cooler temperatures than in conventionally warm spaces (Nelson et al. 1984). Maintain room temperatures in offices at 66–70°F (higher in summer). This temperature range, which is becoming more common in offices, represents the lower end of the ASHRAE comfort envelope. Temperatures lower than this might cause some discomfort because of the sedentary nature of most office work.

Changes to the humidity level in a building affect the sensation of warmth experienced by occupants. The higher the r.h., the lower the air temperature can be to generate an equivalent sensation of warmth in occupants. Combine the recommended temperature of 66–70°F with the recommended r.h. of 35–40 percent for optimal thermal comfort. Make variations as necessary depending on the tasks being performed in the space.

INDIVIDUAL CONTROL

Encourage people to control and adjust their own thermal comfort. Keep thermostat zones small, and locate the thermostat controlling a given zone unmistakably in that zone. Work with occupants to get them to understand that they may use clothing to control their own thermal comfort to some degree and should bring in sweaters and jackets for occasional use.

Engineers who design building systems often recommend that building users not be allowed to use table fans, small radiant heaters, and humidifiers during the work day to keep themselves comfortable. In a properly heated and ventilated building, they argue, these additions should not be necessary, as they place extra demands on the HVAC system. However, engineers cannot account fully for individual differences, for varying office tasks, and for all the heat-generating equipment that might be introduced over the life of a building.

Develop a building policy regarding the addition of fans and similar appliances that is likely to meet the needs of users while respecting the constraints of the physical building. For an office to be truly thermally comfortable to occupants, some individual adjustment of environmental conditions is likely. Women generally feel cooler than men, and sedentary workers feel cooler than active ones. The policy should be upgraded and examined regularly to meet the requirements of new users and changed office arrangements.

ELECTRONIC EQUIPMENT

Extra cooling is required from the air handling systems if VDTs and other electronic equipment are in use. In some work spaces, enclosed areas are filled with electronic equipment that can generate as much as 2,400 BTUs per hour each. Remove the extra heat from the space to make it habitable by workers with an additional system, in much the same way as you would cycle extra air through a conference room if it was in use with a closed door. The requirements of extra heat loads are often badly handled by conventional HVAC system designs. It may be more sensible to disperse heat-generating electronic equipment throughout the building, if this is practical from a work standpoint.

The problem of extra heat removal is best handled before the building is built rather than after the space is in use, when retrofit improvements must be implemented.

HEAT FROM WINDOWS

Solar gain can be expected along the south and west façades of most buildings in the Northern Hemisphere. Manage this in part by exterior or interior window shading. In northern climates as well as in the south, window glass concentrates sunshine and can generate considerable heat even in midwinter.

Avoid placing electronic equipment in offices with a southern exposure (in the Northern Hemisphere). Heat is also emitted from light fixtures; do not locate air supply diffusers over the fixture. However, for efficiency, do locate return air vents over light fixtures because heated air can be drawn out before circulating through the space.

Some heat transfer from sunshine can be controlled by attaching heat-resistant film to the outside of the windows. However, such film also reduces the quality and quantity of daylight admitted by window glass and can make an office interior look gloomy on cloudy days. Select the heat-resistant material that allows the most light transmission. Translucent material that can be lowered or raised like blinds on the interior of the window is preferable in climates where outdoor materials will deteriorate over time.

CONCLUSIONS

These guidelines are clearly all related. It is not possible to plan part of an office space without taking all the other parts into consideration. Therefore, it is impossible to identify those that are more and those that are less important, except in terms of the relative importance of the building-in-use dimensions.

It is also clear that guidelines for different areas of building performance may contradict one another. For example, acoustic partitions may attenuate noise in an open-plan office, but if they are installed in any quantity they may impede air circulation. They protect acoustic privacy better if they extend to the floor; but air delivered at the ceiling circulates better if partitions are a few inches off the floor.

Another example is that ceilings should be hard and smooth to maximize the efficiency of the light emitted from light fixtures, but they should be absorbent and textured to ensure the quality of acoustic conditions in an open space. Windows should be large and plentiful to provide occupants with views and natural light; they are also good sound reflectors, causing sound to carry, and they cause glare and heat gain. In conventional HVAC systems, air can be humidified and supplied at lower temperatures for occupant comfort, but this reduces the actual volume of air supplied. To provide workers with more air to breathe, air must be supplied at warmer temperatures, and to enable people to tolerate the higher temperatures, humidity must be reduced. What a paradox!

At the stage we as a society are presently at in terms of consumer awareness of buildings and of our environment, these contradictions remain part and parcel of the state of the art of modern office design. The technology is there to improve environmental quality in relatively simple ways, but it is not habitually used because the economics and processes of the building industry do not encourage innovation. Most innovative engineering solutions result in a product that is bigger, but not always better, better than any before it. Such engineering innovation often bears little connection to the experience of the human beings who are the "users" of the system. Moreover, relatively little is known about people's behavior in the context of these bigger and better solutions; only recently have experts begun to study systematically the behavior of people in buildings over time.

The development and construction industry as we know it does not lend itself to innovation and evaluation. There is no built-in quality control for the production of buildings. Technological innovation is randomly applied, and feedback is not systematically incorporated into the information flow through the industry. Therefore, the initiative for quality control must come from building users and managers. These interest groups really only emerge once a building is built, and it is their interests and concerns that are served by these guidelines.

Although the guidelines for office improvement contradict each other in places, they do emphasize the common and easy-to-remedy errors of office planning. They synthesize what is known, and they try to speculate on what is not yet known about people's behavior in buildings. They address the space-planning decisions made by office occupants and building managers on a regular basis as part of changing and retrofitting existing offices as well as the design of new buildings. So many office buildings have been built during the last decade in major urban centers that the need to improve existing buildings is now as strong as the need to design new ones.

Now is the time for owners and managers to start measuring office-building quality. Use the building-in-use assessment system, generate building-in-use profiles and norms, and compute the index for the seven dimensions. Then priorities can be set on environmental improvements. And with the budget thereby acquired, use the guidelines to improve buildings, and with them, office worker morale and productivity. With a properly implemented assessment system, this is a no-lose proposition.

In the world we live in, the difference between a good and a bad building is a matter of degree. It is not too soon to start creating work environments that are technologically and aesthetically innovative, offices that are truly interactive with workers and incorporate the innovative technology conventionally known as the "office of the future." The place to start is the building we work in; the time to start is now.

APPENDIX
TECHNICAL SUMMARY: STATISTICAL APPROACH TO DATA ANALYSIS*

The survey questionnaire database contains responses to approximately 2,900 questionnaires, not all of which were completed or, therefore, used. Each questionnaire contained thirty-five rating scales designed to measure ambient environmental conditions. On each scale, respondents assigned a rating from 1 (bad or uncomfortable) to 5 (good or comfortable). The three midpoints on the scales were not labeled, but the labels at 1 and 5 were designed to indicate that 3 is a midpoint between comfortable and uncomfortable, with 2 and 4 as intermediate values.

The approach to statistical analysis included the following steps. First, the thirty-five environmental rating scales were submitted to a factor analysis. Seven factors emerged consistently using a variety of different factor-analysis procedures. Second, simplified scores on the seven factors were obtained for each respondent by taking the means of the values on the

* Condensed from *The Building-in-Use Assessment Methodology* vols. 1 and 2 by Richard Dillon and Jacqueline Vischer. Ottawa, Canada: Public Works Canada, 1988.

rating scales that were important in each factor. Another series of tests was then used to determine differences among the buildings on each of the factors. Analysis of variance, multivariate analysis of variance, and discriminant analysis were used to see which building differences were significant and to interpret the differences in terms of the factors. Finally, factors were then used to predict such global variables as work-station satisfaction, whether the work environment helped or hindered people doing their work, and health complaints, using multiple regression techniques and canonical correlation. Each of these steps is described below.

FACTOR ANALYSIS

A series of exploratory factor analyses was performed to summarize patterns of intercorrelations among ratings of environmental quality of the buildings' interiors. In this analytic technique, a large number of variables (in this case, thirty-five scale ratings) is reduced to a smaller number of unobservable dimensions or constructs called *factors* that can be interpreted and used for further analysis. In factor analysis, the assumption is made that questions that are highly intercorrelated are measuring related aspects of the same dimension. Although this dimension cannot be observed, analysis of the questions that determine the dimension can provide insights into the nature of the dimension.

Six analyses that combined three factor-analysis methods with two popular rotation techniques were performed. *Principal component factor analysis, iterated principal factor analysis,* and *maximum likelihood factor analysis* were all performed and their factors rotated with varimax and equamax rotations. Varimax and equamax rotations result in factors that are orthogonal; thus, the factors obtained are treated as if they are independent. The fact that the same factors emerged, with only minor variations, regardless of analytical method or rotation type, is to be expected with a large number of variables (35) and large sample size (1,685 useable questionnaires out of a total of 2,900). The proportions of variance and factor loadings reported are those obtained from the *maximum likelihood factor analysis* with *varimax* rotation.

The factors identified were labeled Air Quality (21 percent), Noise Control (7 percent), Thermal Comfort (4 percent), Spatial Comfort (4 percent), Privacy (3 percent), Lighting Comfort (3 percent), and Building Noise Control (2 percent). These are the building-in-use dimensions described in chapter 6. One of the objectives of factor analysis is to account for as large a proportion of variance in the original thirty-five scales with as few factors as possible. The values in parentheses are the proportions of variance in the original thirty-five scales accounted for by each factor. For example, 21 percent of the variance is determined by Air Quality, whereas only 2 percent is determined by Building Noise Control. Since additional factors account for progressively less variance, the inclusion of additional factors would be of little value. With only seven factors, it is possible to account for 45 percent of the variance in the original scales.

The component scales for each factor with varimax rotation loadings above 0.30 are listed in table A-1. Only scales with loadings above 0.30 are presented in this table as these are the correlations that are important enough to enable us to interpret the factor.

To validate the factors, separate factor analyses were performed on each building. Essentially the same factors emerged with minor variations. The Air Quality factor emerges clearly in every building. Lighting Comfort and Spatial Comfort show some variation by building, as Lighting Comfort is sometimes separated from a Daylighting factor, and Privacy is sometimes included under Spatial Comfort.

SIMPLIFIED FACTORS

Simplified factor scores were constructed for each respondent to enable the questionnaire to be reduced to its essential scales.

Initially, two sets of simplified factor scores were compared. The first set used the means of the scales presented in table A-1 as a definition of a simplified factor. For example, Air

TABLE A-1
LIST OF SCALES THAT MAKE UP EACH FACTOR OR BUILDING-IN-USE DIMENSION.

Factor	Scale	Rotation Loading
Thermal Comfort	Cold Temperatures	0.77
	Temperature Comfort	0.61
	Temperature Shifts	0.60
	Draftiness	0.58
Privacy	Voice Privacy	0.80
	Telephone Privacy	0.73
	Visual Privacy	0.50
Noise Control	General Noise Levels	0.87
	Specific Office Noises	0.82
	Noise Distractions	0.75
Spatial Comfort	Amount of Space in Work space	0.65
	Work Storage	0.65
	Furniture Arrangement	0.59
	Personal Storage	0.56
	Furniture Comfort	0.44
	Visual Access to Work space	0.33
Lighting Comfort	Electric Lighting Comfort	0.67
	Glare from Lights	0.66
	Too Bright	0.55
	Colors	0.47
	Daylighting	0.38
Building Noise Control	Noise from the Lights	0.68
	Noise from the Air Systems	0.55
	Noise from Outside	0.55
Air Quality	Air Movement	0.88
	Air Freshness	0.85
	Ventilation Comfort	0.79
	Odors	0.50
	Humidity/Dry Air	0.47
	Warm Temperatures	0.40

TABLE A-2
CORRELATIONS AMONG FACTOR SCORES

Factor	Correlation Coefficient
Thermal Comfort	0.96
Privacy	1.00
Noise Control	1.00
Spatial Comfort	0.98
Lighting Comfort	0.87
Building Noise Control	0.91
Air Quality	0.92

Quality scores were obtained for each respondent by summing the scores for Air Movement, Air Freshness, Ventilation Comfort, Odors, Humidity, and Warm Temperatures and dividing by six. Thus, the simplified factor scores treat all scales with loading above 0.30 as equally important and all scales with loadings at or below 0.30 as unimportant.

The first set of simplified factor scores used thirty of the thirty-five scales available. To reduce the number of scales even further, a second set of simplified factors was defined. In this set, the criterion for inclusion in a factor was changed from a loading of at least 0.30 to a proportion of variance of at least 0.25. Since loadings are correlations, the squares of the loadings show the proportions of variance shared by scales and factors. A criterion of 0.25 as a minimum proportion of variance is equivalent to a minimum factor loading of 0.50. Correlations between factors on the large set and factors on the minimal set are shown in table A-2. These coefficients indicate that the large set and the minimal set are highly correlated. Little is therefore lost by using the set based on fewer scales.

In table A-3, the means and standard deviations for the large set and the minimal set on each factor are reported; standard deviations are essentially equivalent, especially for the minimal-set factors. In this table, the factors are listed in descending order of their values, or means, as computed by the two procedures. The Building Noise Control factor is slightly restricted in variability by the proximity of the mean to the upper

TABLE A-3
MEANS AND STANDARD DEVIATIONS OF THE FACTORS AS COMPUTED BY THE TWO SIMPLIFICATION PROCEDURES

	Large Set		Minimal Set	
Factor	**Mean**	**s.d.**	**Mean**	**s.d.**
Building Noise	4.3	0.68	4.4	0.74
Spatial Comfort	3.3	0.85	3.3	0.94
Lighting Comfort	3.2	0.88	3.3	1.00
Thermal Comfort	2.9	1.00	2.8	1.02
Noise Intrusion	2.9	1.13	2.9	1.13
Air Quality	2.6	0.96	2.3	1.11
Privacy	2.3	1.08	2.3	1.08

endpoint of the scale. The high-scoring factors are those building environment factors that perform best in office buildings, and the low-scoring factors are those that perform worst or are potentially the greatest problem in office buildings.

ANALYSIS OF VARIANCE, MULTIVARIATE ANALYSIS OF VARIANCE, AND DISCRIMINANT ANALYSIS

A series of analyses of variance were performed to determine significant variation on the seven factors among the five buildings. This procedure compared the means on each factor across the five buildings and determined whether or not there was enough variation between any of the building means to indicate a significant difference, or if the amount of variation in means could best be attributed to chance fluctuation.

Table A-4 shows the sample sizes, F-scores, probability, and proportion of variance associated with each simplified factor.

The fact that there is not a lot of variation in these factors among buildings, together with the large differences between factors, is indicative of the utility of the factors as real constructs underlying occupants' patterns of response to interior environmental quality.

TABLE A-4
RESULTS OF ANALYSIS OF VARIANCE SHOWING SIGNIFICANT DIFFERENCES ACROSS BUILDINGS ON THE BUILDING-IN-USE DIMENSIONS

Factor	Sample Size*	F	p**	R^2
Thermal Comfort	2172	6.16	.0001	.01
Privacy	2120	10.32	.0001	.02
Noise Control	2348	143.18	.0001	.20
Spatial Comfort	2349	4.23	.002	.01
Lighting Comfort	2284	10.19	.0001	.02
Building Noise Control	2380	154.57	.0001	.21
Air Quality	2319	11.41	.0001	.02

* Sample size varies according to amount of missing data in each case.
** Although there are highly significant differences between the buildings on all seven simplified factors, these differences are small. They account for only 1 to 2 percent in the variance of all factor scores except Noise Control (20 percent), and Building Noise Control (21 percent). With the extremely large sample sizes, statistical significance is misleading: proportion of variance accounted for is a more appropriate index of importance.

MULTIPLE REGRESSION AND CANONICAL CORRELATION

The seven simplified factors were used to construct a model to predict three criterion variables. The first variable, *work-station satisfaction* (Mean = 3.2, s.d. = 0.95), provides a global rating of occupants' satisfaction with their work environments, specifically, the work station or microenvironment in which they spend most of their time. The second variable, *workability* (Mean = 3.3, s.d. = 1.02), describes occupants' ratings of how much the immediate office environment helps them or hinders them in doing their work. The third global variable is occupants' *health status* in the building. Each of the global variables is scored on a scale of 1 to 5, except for Health Status, which is an index with values from 1 through 4 constructed from a series of fourteen building-related symptoms of ill-health to which respondents had given "yes" or "no" responses.

A least squares regression analysis was carried out separately for each criterion variable using the seven building factors as predictors. Multiple regression is a statistical procedure for

TABLE A-5
RESULTS OF REGRESSION ANALYSIS OF PREDICTABILITY OF THREE GLOBAL VARIABLES BY THE SEVEN BUILDING-IN-USE DIMENSIONS

	Large Set			Minimal Set		
Criterion Variable	R^2	F	p	R^2	F	p
Satisfaction	.24	70.14	.0001	.22	67.95	.0001
Workability	.40	150.63	.0001	.40	157.39	.0001
Health Status	.21	67.02	.0001	——	——	——

determining the best combination of variables to predict the criterion variables. This was followed by canonical correlation analyses, in which all three dependent variables were predicted simultaneously.

Table A-5 summarizes R^2s, Fs, and the probabilities associated with Fs, using the large set and the minimal set of simplified factor scores to predict the three criterion variables. It is clear from table A-5 that the minimal-set factors are essentially equivalent to the large-set factors for predicting the criterion variables. The results presented in chapter 6 are therefore based on analyses of the minimal-set factors.

The multiple correlations reported above examine the predictability of the three criterion variables separately. Canonical correlation analysis examines the predictability the three criterion variables simultaneously. The analysis determines the linear combination of factors that does the best job of predicting the most complete combination of the criterion variables. The canonical correlation is the correlation between scores on these linear combinations. The purpose of canonical analysis is to account for a larger proportion of the variance in the criterion variables than can be predicted for each separately.

The analysis found that the squared canonical correlation accounts for 43 percent of the variance in the criterion variables. The largest proportion of variance in any one outcome variable is 40 percent. Therefore, the canonical correlation analysis does not do much better than the multiple regression of each global variable separately. In short, for these data, the canonical analysis provides the same information as the three separate regressions.

BUILDING-IN-USE ASSESSMENT

Because the seven factors are consistent across all buildings, the scores computed for each of them can be construed as *norms* for the factors, and other building-occupant survey results can be compared to them. This is the basis for the building-in-use assessment system.

The generic questionnaire used for building-in-use assessment comprises the scales that make up the minimal set factors. It is presented at the end of this section. The generic questionnaire is a tested, reliable instrument for measuring people's attitudes about the office environments in which they work. Each of the scales used in the questionnaire has been tested and is a component of a cluster or factor representing building users' responses to their office environment. The number of scales is adequate to predict dimensions of building users' environmental experience. Each factor or environmental dimension has a normative rating or score.

The score for each factor (the norm for each building-in-use dimension) computed on the minimum number of com-

TABLE A-6
NORMATIVE SCORES AND STANDARD DEVIATIONS FOR THE SEVEN BUILDING-IN-USE DIMENSIONS

Factor	Norm	s.d.
Building Noise Control	4.4	0.74
Spatial Comfort	3.3	0.94
Lighting Comfort	3.3	1.00
Noise Control	2.8	1.13
Thermal Comfort	2.9	1.02
Air Quality	2.3	1.11
Privacy	2.3	1.08

TABLE A-7
LIST OF SCALES FROM WHICH EACH FACTOR SCORE IS COMPUTED

Factor Score	=	Sum of Component Scales	Divided by:
Thermal Comfort	=	Temperature comfort + Cold temperatures + Temperature shifts	3
Privacy	=	Visual privacy + Voice Privacy + Telephone privacy	3
Noise Control	=	Noise Distractions + General Noise Level + Specific Office Noises	
Spatial Comfort	=	Furniture Arrangement + Amount of Space + Work storage + Personal storage	4
Lighting Comfort	=	Electric Lighting + Too Bright + Glare from lights	3
Building Noise Control	=	Air systems noise + Lights noise + Noise from outside	3
Air Quality	=	Ventilation + Air freshness + Air movement	3

ponent scales is provided in table A-6. These factor scores are arranged in descending order. They approximate a population of possible factor scores against which new scores can be compared. The scales from which each factor is computed for the assessment of environmental quality are listed in table A-7.

It must be emphasized that building-in-use assessment does not evaluate whether performance of a building on the factors is categorically good or bad, right or wrong. Rather, it allows comparison between one building and a baseline of typical office buildings to determine whether the building's performance on each dimension is consistent with, better than, or worse than that of known buildings.

Using standard, small-sample, inferential statistical techniques, with the values of the factors (see table A-7) serving as population means and standard deviations, it is possible to calculate confidence intervals around the means for each factor. According to the central limit theorem, for reasonably small sample sizes, the population (and, therefore, a sample) does not have to have normally distributed factor scores. Moreover, the sample size required is independent of the number of occupants in any building.

With randomly selected samples of not less than thirty respondents, it is possible to make a relative assessment of any building on the seven factors. Tables can be constructed according to the confidence levels set at the outset of the study (for example, 90 percent and 99 percent), which show the significance of deviation on either side of the norm.

To use these tables, the scale ratings for all respondents are transformed into factor scores, and the mean for each factor is calculated. The table for the sample size closest to actual sample size is selected, and the seven obtained sample means are entered on this table. This table then represents the profile for the building relative to the normative data. As explained in chapter 7, if a mean for any factor falls within the 90 percent confidence limits, this indicates that the factor is within the normative range. A mean that falls between the 90 and 99 percent confidence limits is an indicator of possible concern. A mean that falls beyond the 99 percent confidence limits is of serious interest and indicates the need for a more detailed analysis of the building using techniques of building-performance testing.

GENERIC QUESTIONNAIRE

PLEASE RANK THE FOLLOWING ATTRIBUTES OF YOUR PARTICULAR DESK LOCATION IN THIS BUILDING. PLEASE CIRCLE THE APPROPRIATE NUMBER BETWEEN 5 (COMFORTABLE) AND 1 (UNCOMFORTABLE) THAT BEST SUMMARIZES YOUR EXPERIENCE OF WORKING HERE:

Attribute	1	2	3	4	5
Temperature Comfort	1 Bad	2	3	4	5 Good
How Cold It Gets	1 Too Cold	2	3	4	5 Comfortable
Temperature Shifts	1 Too Frequent	2	3	4	5 Generally Constant
Ventilation Comfort	1 Bad	2	3	4	5 Good
Air Freshness	1 Stale	2	3	4	5 Fresh
Air Movement	1 Stuffy	2	3	4	5 Circulating
Noise Distractions ...	1 Bad	2	3	4	5 Good
General Office Noise Level (Conversation and Equipment)	1 Too Noisy	2	3	4	5 Comfortable
Specific Office Noises (Voices and Equipment)	1 Disturbing	2	3	4	5 Not a Problem
Voice Privacy at Your Desk	1 Bad	2	3	4	5 Good
Telephone Privacy at Your Desk	1 Bad	2	3	4	5 Good
Noise from the Air Systems	1 Disturbing	2	3	4	5 Not a Problem
Noise from the Office Lighting	1 Buzz/Noisy	2	3	4	5 Not a Problem
Noise from Outside the Building	1 Disturbing	2	3	4	5 Not a Problem

GENERIC QUESTIONNAIRE (*Continued*)

Furniture Arrangement in Your Work Space	1	2	3	4	5
	Bad				Good
Amount of Space in Your Work Space	1	2	3	4	5
	Bad				Good
Work Storage	1	2	3	4	5
	Insufficient				Adequate
Personal Storage	1	2	3	4	5
	Insufficient				Adequate
Visual Privacy at Your Desk	1	2	3	4	5
	Bad				Good
Electric Lighting	1	2	3	4	5
	Bad				Good
How Bright Lights Are	1	2	3	4	5
	Too Much Light				Does Not Get Too Bright
Glare from Lights	1	2	3	4	5
	High Glare				No Glare

BIBLIOGRAPHY

Allen, Tom J. 1977. *Managing the flow of technology.* Cambridge: MIT Press.

American Productivity Center. 1982. *White collar productivity: The national challenge.* Grand Rapids: Steelcase Inc.

Archea, John. 1977. The place of architectural factors in behavioral theories of privacy. *Journal of Social Issues* 33(3):116–37.

ASHRAE. 1981. *Ventilation for acceptable indoor air quality.* Atlanta: ASHRAE.

Barge, F. A. n. d. "Speech privacy in open plan office environments." Pittsburgh: Westinghouse R&D Center. Mimeo.

Barker, Roger. 1968. *Ecological psychology: Concepts and methods for studying the environment for human behavior.* Stanford, California: Stanford University Press.

Barker, Roger, and Schoggin, P. 1973. *Qualities of community life.* San Francisco: Jossey-Bass.

Becker, Franklin C. 1981. *Workspace: Creating environments in organizations.* New York: Praeger.

———, et al. 1986. *ORBIT 2.* Stanford, Connecticut: Harbinger Group.

Bikson, T. K., and Gutek, B. A. 1983. *Advanced office systems: An empirical look at utilization and satisfaction.* Santa Monica: Rand Corporation.

Birch, David. 1986. *America's office needs 1985–1995.* Chicago: Research Report for MIT Center for Real Estate Development and Arthur Anderson and Co.

Bon, Ranko. 1986. Timing of space: Some thoughts on building economics. *Habitat International* 10(4):101–7.

BOSTI 1982. *The impact of the office environment on productivity and the quality of working life.* Buffalo: Westinghouse Furniture Systems.

Boston Globe. 30 June, 1987.

Boyce, P. R. 1974. Users' assessments of a landscaped office. *Journal of Architectural Research* 3:44–62.

———. 1975. "The luminous environment." In *Environmental interaction: Psychological approaches to our physical surroundings,* edited by D. Canter and P. Stringer. New York: International Universities Press.

———. 1981. *Human factors in lighting.* New York: Macmillan.

Brandon, Peter S. 1984. "Cost versus quality: A zero sum game?" In *Designing for building utilization,* edited by J. A. Powell, I. Cooper, and S. Lera. New York and London: E. and F. N. Spon.

Brill, M., with Margulis, S. T., and Konar, E. 1985. *Using office design to increase productivity* (2 vols.). Buffalo: BOSTI and Westinghouse Furniture Systems Inc.

Brooks, M. J., and Kaplan, A. 1972. The office environment: Space planning and effective behavior. *Human Factors* 14(5):373–91.

Bryan, Harvey. 1983. Development of an integrated daylighting design methodology, in *Proceedings of International Daylighting Conference,* Phoenix, Arizona. edited by Thomas Vonier.

Bryan, H., Clark, G., and Vischer, J. 1988. The effects of variability in lighting. Paper presented at Advanced Comfort Systems Symposium, Troy, New York; May 2–6.

Cakir, A., Hart, D. J., Stewart, T. F. M. 1980. *Visual display terminals: a manual covering workplace design, ergonomics, health and safety.* New York: John Wiley and Son.

Canter, David. 1975. "Buildings in use." In *Environmental interaction: Psychological approaches to our physical surroundings,* edited by D. Canter and P. Stringer. New York: International Universities Press.

Carlton-Foss, John. 1983. The tight building syndrome: Diagnosis and cure. *ASHRAE Journal* (December):38–41.

The Citizen (Ottawa, Canada). 16 Nov. 1979; 11 March 1980; 8 April 1981; 10 Nov. 1982; 31 July 1984; 1 Feb. 1985; 13 March 1985.

Coe, James. 1983. Work posture, workstation layout and design in the electronic office. Melbourne, Australia:Centre for Applied Ergonomics, Royal Melbourne Institute of Technology.

Cohen, Aaron, and Cohen, Elaine. 1983. *Planning the Electronic Office*. New York: McGraw-Hill.

Craig, Marianne. 1981. *Office workers' survival handbook: A guide to fighting health hazards in the office.* London: BSSRS Publications.

Craik, Kenneth, and Zube, Ervin. 1976. *Perceiving environmental quality: research and applications.* New York: Plenum Press.

Davis, Gerald, and Szigeti, F. 1986. "Planning and programming offices: Determining user requirements." In *Behavioral issues in office design,* edited by Jean Wineman. New York: Van Nostrand Reinhold.

de Zeeuw, Gerard. 1980. "Methodological problems of psychotherapy research." In *Psychotherapy: Research and training,* edited by W. de Moor and H. R. Wijngaarden. New York: Elsevier/North Holland Biomedical Press.

______ . 1981. "Systems and change: Some concepts." University of Amsterdam. Mimeo.

Dillon, Richard, and Vischer, Jacqueline C. 1988. *The Building-in-use assessment methodology.* Ottawa, Canada: Public Works Canada. 2 vols.

Dolden, M., and Ward, R., eds. 1986. *The impact of the work environment on productivity: Proceedings of a workshop.* Washington, D. C.: National Science Foundation and Architectural Research Centers Consortium.

Duffy, Francis. 1983. *Information technology and office design: The ORBIT study.* London: DEGW and EOSYS Ltd.

Ekuan, Shoji, 1985. "Office environments in transition." In *The impact of the work environment on productivity: Workshop proceedings,* a proposal by Ward et al. to the National Science Foundation, Civil and Environmental Engineering.

Ellis, Peter, ed. 1986. *Achieving office quality: A report of international research on the nature of office quality and its relationship to the processes of development, design, procurement and management.* London and Bonn: Anglo-German Foundation.

———. 1985. Office productivity. *Facilities* 3:13–16.

Evans, G., ed. 1982. *Environmental stress.* New York: Cambridge University Press.

Facilities Planning News. June 1986.

Fanger, P. O. 1972. *Thermal comfort.* New York: McGraw-Hill.

Finnegan, M. J., Pickering, C. A., and Burge, P. S. 1984. The sick building syndrome: Prevalence studies. *British Medical Journal* 289:1573–75.

Flynn, John E., Kingsbury, Howard F., and Gillette, Gary. 1979. *A review of the state of the art in daylighting in buildings.* Pennsylvania State University: Department of Architectural Engineering.

Foucault, Michel. 1977. *Discipline and punish: The birth of the prison* (English translation by Alan Sheridan). New York: Pantheon Books.

Friedmann, Arnold, Zimring, Craig, and Zube, Ervin. 1978. *Environmental design evaluation.* New York: Plenum Press.

Goldman, Ralph. 1978. Assessment of thermal comfort. *ASHRAE Transactions* 84 (1):71–95.

Goodrich, R. J. 1982. "The perceived office: The office environment as experienced by its users." In *Behavioral issues in office design,* edited by Jean Wineman. New York: Van Nostrand Reinhold.

Gray, John M., Daish, John R., and Kernohan, David Q. 1986. "A touring interview method of building evaluation: The place of evaluation in building rehabilitation." In *Building performance: Function, preservation, rehabilitation ASTM STP 901,* edited by G. Davis. Philadelphia: American Society for Testing and Materials.

Gregory, Judith. 1983. "The electronic sweatshop." In *Perspectives on women in the 1980's,* edited by Joan Turner and Lois Emery. Winnipeg: University of Manitoba Press.

Harris, David A., Palmer, Alvin E., Lewis, M. S., Munson, D. L., Meckler, G., and Gerdes, R. 1981. *Planning and designing the office environment.* New York: Van Nostrand Reinhold.

Harris, L. and Assoc. 1978, 1980. *The Steelcase national study of office environments: Do they work?* (2 vols). Grand Rapids: Steelcase Inc.

Hawkins, L. H. 1984. "The possible benefit of negative-ion generators." In *The Health Hazards of VDT's,* edited by B. G. Pearce; London: John Wiley.

Hedge, Alan. 1982. The open plan office: A systematic investigation of employee reactions to their work environment. *Environment and Behavior* 14(5):519–42.

Heerwagen, Judith H., and Heerwagen, Dean R. 1984. Energy and psychology: Designing for a state of mind. *Journal of Architectural Education* 37 (3 and 4):34–37.

Herzberg, Frederick. 1966. *Work and the nature of man.* Cleveland: World Publishing Co.

Hillier, Bill, and Hanson, Julienne. 1984. *The social logic of space.* Cambridge: Cambridge University Press.

Hopkinson, R. G. 1963. *Architectural physics: Lighting.* London: H. M. Stationery office.

Kleeman, Walter. 1986. "The office of the future." In *Behavioral issues in office design*, edited by Jean Wineman. New York: Van Nostrand Reinhold.

Lam, William M. C. 1977. *Perception and lighting as form-givers for architecture.* New York: McGraw-Hill.

Levin, Hal. 1985. A report on indoor pollution research and its potential and actual applications in architectural practice. *Proceedings of research and design 85.* Washington DC: American Institute of Architects.

______ . 1986. "Indoor air pollution research and its applications in office building development and operation." In *The changing office workplace*, edited by J. T. Black, K. Roark, and L. S. Schwartz. Washington, D. C.: Urban Land Institute and Building Owners and Managers Association.

Lindheim, Roslyn. 1974. Environments for the elderly: Future-oriented design for living. Paper presented to the American Association for the Advancement of Science, San Francisco, California.

Lorimer, James. 1979. *The developers.* Toronto: J. Lorimer and Co.

Lueder, Rani. 1986. *The ergonomics payoff: Designing the electronic office.* New York: Nichols Publishing Co.

MacFarlane, W. V. 1978. Thermal comfort studies since 1958. *Architectural Science Review* 21(4):89–92.

Makower, Joel. 1981. *Office hazards: How your job can make you sick.* Washington, D. C.: Tilden Press.

Manning, Peter. 1968. Lighting in relation to other components of the total environment. *Transactions of the Illuminating Engineering Society* 33(4):159–66.

Marans, Robert, and Spreckelmeyer, Kent F. 1981. *Evaluating built environments: A behavioral approach.* Ann Arbor: Institute for Social Research and Architectural Research Laboratory, University of Michigan.

______ . 1982. "Perceived quality of residential environments: Some methodological issues." In Craik and Zube, *op. cit.*

______ , and Spreckelmeyer, Kent F. 1982. Measuring overall architectural quality: A component of building evaluation. *Environment and behavior* 14(6):652–70.

Margolis, Stephen. 1981. *A Methodology for Evaluating Housing-In-Use: A Case Study Approach*. Washington D. C.: US Dept. of Commerce, National Bureau of Standards.

Mill, P. 1984. *Transdisciplinary building diagnostics: total building Performance.* Ottawa, Canada: Public Works Canada, Architectural and Building Sciences.

Miller, Hugh. 1985. Diagnostic methods for evaluating conditions in existing buildings. Proceedings of the International Conference on building use and safety Technology: 29–34.

Moleski, W. H., and Lang, J. T. 1982. "Organizational goals and human needs in office planning." In *Behavioral issues in office design,* edited by Jean Wineman. New York: Van Nostrand Reinhold.

Murrell, K. F. H. 1969. *Ergonomics: Man and his working environment.* London: Chapman and Hall.

National Research Council. 1985. *Building diagnostics: A conceptual framework.* Washington, D. C.: National Academy Press.

Nelson, T. M., Nilsson, T. H., and Johnson, M. 1984. Interaction of temperature, illumination and apparent time on sedentary work fatigue. *Ergonomics* 27(1):89–101.

Nemecek, J., and Grandjean, E. 1973. Results of an ergonomic investigation of large-scale offices. *Human Factors* 15(2):111–24.

Newsweek. 1985. Beware sick building syndrome. 7 January:58–60.

Oldham, G., and Brass, D. 1979. Employee reactions to an open plan office: A naturally occurring quasi-experiment. *Administrative Science Quarterly* 24:267–84.

Passini, Romedi. 1984. *Wayfinding in architecture.* New York: Van Nostrand Reinhold.

Pirsig, Robert M. 1975. *Zen and the art of motorcycle maintenance: An inquiry into values.* New York: Bantam Books.

Planas, Rod E. 1978. Perfect open plan priority: The human element. *Buildings* (March):74–75.

Public Works Canada. 1985. *Stage one total building performance* (13 vols). Ottawa: Public Works Canada, Division of Architectural and Building Sciences.

______. 1986. *Report to national capital region.* Ottawa: Public Works Canada, Division of Architectural and Building Sciences.

Rand. George. 1986. Whatever happened to the office of the future? *Architecture* (AIA) 75(12):106–8.

Rea, Mark. 1982. Calibration of subjective scaling responses. *Lighting Research and Technology* 14(3):121–29.

Rohles, Frederick H. and Jones, Byron W. 1983. A fan in winter. *Proceedings of the Human Factors Society 27th Annual Meeting:* 742–745.

Rubin, Arthur. 1984. *The automated office—An environment for productive work, or an information factory?* Washington, D. C.: National Bureau of Standards, Center for Building Technology, National Engineering Laboratory.

Rush, Richard, ed. 1986. *The building systems integration handbook.* New York: John Wiley and Sons.

Seiler, John A. 1984. Architecture at work. *Harvard Business Review* (Sept.-Oct.):120.

Shelly, M. 1969. *Analyses of satisfaction vol. 1.* New York: MSS Educational Publishing Co.

Sodergren, David, and Puntilla, Antero. 1983. "A CO_2-controlled ventilation system: Pilot study." Stockholm: Swedish Council for Building Research. Mimeo.

Sommer, Robert. 1983. *Social design.* Englewood Cliffs, New Jersey: Prentice-Hall.

Spreckelmeyer, Kent F. 1985. "Environmental norms in the workplace." University of Kansas. Mimeo.

Steelcase. 1983. *The impact of open office furniture systems on employee productivity: Three case studies.* Grand Rapids: Steelcase Inc.

______ . 1987. *The office environment index: 1987 summary report.* by L. Harris and Assoc. Grand Rapids: Steelcase Inc.

Steele, F. I. 1973. *Physical settings and organizational development.* Reading: Addison-Wesley Publishing Co.

Stellman, Jean. 1977. *Womens work, womens health.* New York: Pantheon Books.

______ , et al. 1982. *Epidemiological study of office buildings for building health problems.* Ottawa: Public Works Canada, Columbia School of Public Health.

Stone, Philip, and Luchetti, Robert. 1985. Your office is where you are. *Harvard Business Review* (March-April):102–17.

Stramler, C. S., Kleiss, J. A., and Howell, W. C. 1983. Thermal sensation shifts induced by physical and psychological means. *Journal of Applied Psychology* 68(1):187–93.

Sundstrom, Eric. 1986. *Workplaces: The psychology of the physical environment in offices and factories.* New York: Cambridge University Press.

______ , Burt, Robert E., and Kamp, D. 1980. Privacy at work: Architectural correlates of job satisfaction and job performance. *Academy of Management Journal* 23(1):101–17.

Szilagyi, A., and Holland, W. 1980. Changes in social density: Relationships with functional interaction and perceptions of job characteristics, role stress and work satisfaction. *Journal of Applied Psychology* 65(1):28–33.

Taylor, F. W. 1911. *The Principles of Scientific Management.* New York: Harper.

Turner, William A., and Bearg, David. 1987. Identifying and avoiding air quality problems. *Heating, Piping, Air Conditioning* (February):45–49.

USA Today. 14 April 1986.

Ventre, Francis. 1988. Sampling building performance. Paper presented at Facilities 2000 Symposium, Grand Rapids, Michigan, April 24–26.

______. 1982. "Analysis of the problem of studying people in buildings: How to de-compose buildings-in-use." In *The problems of actors and actions,* edited by Annette Pedretti. Zurich: Princelet Editions.

______. 1985. The adaptation and control model of user needs: A new direction for housing research. *Journal of Environmental Psychology* 5:287–96.

______ 1986a. Post-occupancy evaluation: Tool for designers and users. *Contract* (August):88–89.

______, and Cooper Marcus, Clare. 1986b. Evaluating evaluation: Analysis of a housing design awards program. *Places* 3(1):66–84.

______. 1987a. Office pollution. *Boston Woman* (February):13–15.

______. 1987b. The psychology of daylighting. *Architecture* (June):109–111.

Vonier, Thomas. 1983a. Building diagnostics: Preventive medicine. *Progressive Architecture* 3:143–49.

______. ed. 1983b. *General proceedings: 1983 international daylighting conference.* 16–18 February, Phoenix, Arizona.

Ward, Robertson, et al. 1984. Proposal for "A workshop on the impact of the work environment on productivity" to the National Science Foundation, Civil and Environmental Engineering.

Warnock, A. C. C., Henning, D., and Northwood, T. D. 1972. *Acoustic survey of an open plan landscaped office.* Ottawa: National Research Council, Division of Building Research.

Wells, Brian. 1965. Subjective responses to the lighting installations in a modern office building and their design implications. *Building Science* 1:57–68.

Wetherill, E. A. 1985. Acoustics: The forgotten dimension. Paper presented at Convention Center Management Conference, Tucson, Arizona, 1–4 November.

Wilson, F. 1984. *Building materials evaluation handbook.* New York: Van Nostrand Reinhold.

Wineman, Jean D. 1982. Office design and evaluation: An overview. *Environment and Behavior* 14(3):271–98.

Wyon, Daniel P. 1988. Energy conservation and user control in advanced comfort systems: Work place HVAC versus the computerised technological fix. Paper presented at the Advanced Comfort Systems Symposium, Troy, New York.

Zeisel, John. 1985. "Building purpose: The key to measuring building effectiveness." In Dolden and Ward, *op. cit.*

INDEX